当代建筑师系列

李兴钢
LI XINGGANG

李兴钢工作室　编著

中国建筑工业出版社

图书在版编目(CIP)数据

李兴钢/李兴钢工作室编著. —北京：中国建筑工业出版社，2012.6
（当代建筑师系列）
ISBN 978-7-112-14304-7

Ⅰ.①李… Ⅱ.①李… Ⅲ.①建筑设计-作品集-中国-现代②建筑艺术-作品-评论-中国-现代 Ⅳ.① TU206 ② TU-862

中国版本图书馆CIP数据核字（2012）第091504号

整体策划：陆新之
责任编辑：刘 丹 徐 冉
责任设计：董建平
责任校对：张 颖 陈晶晶

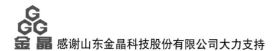感谢山东金晶科技股份有限公司大力支持

当代建筑师系列

李兴钢

李兴钢工作室 编著

*

中国建筑工业出版社出版、发行（北京西郊百万庄）
各地新华书店、建筑书店经销
北京嘉泰利德公司制版
北京顺诚彩色印刷有限公司印刷

*

开本：965×1270毫米 1/16 印张：11¼ 字数：314千字
2012年8月第一版 2012年8月第一次印刷
定价：98.00元
ISBN 978-7-112-14304-7
（22363）

版权所有 翻印必究
如有印装质量问题，可寄本社退换
（邮政编码 100037）

目 录   Contents

李兴钢印象	4	Portrait
建筑的发现与呈现	6	The Discovery and Presentation of Architecture
兴涛接待展示中心	10	Xingtao Reception & Exhibition Center
建川镜鉴博物馆暨汶川地震纪念馆	20	Jianchuan Mirror Museum & Wenchuan Earthquake Memorial
复兴路乙 59-1 号改造	34	Reconstruction of No.B-59-1, Fuxing Road
Hiland·名座	48	Hiland · Mingzuo
纸砖房	58	Paper-brick House
李兴钢工作室室内设计	68	Interior Design of Atelier Li Xinggang
北京地铁 4 号线及大兴线地面出入口及附属设施	80	Accesses and Subsidiaries of Line 4 & Daxing Line of Beijing Subway
北京地铁昌平线西二旗站	90	Xi'erqi Station of Changping Line of Beijing Subway
商丘博物馆	100	Shangqiu Museum
"第三空间"	106	"The Third Space"
元上都遗址博物馆	112	Museum for Site of XANADU
海南国际会展中心	118	Hainan International Conference & Exhibition Center
绩溪博物馆	132	Jixi Museum
元上都遗址工作站	138	Entrance for Site of XANADU
西柏坡华润希望小镇	150	Xibaipo China Resources Hope Town
李兴钢访谈	160	Interview
作品年表	168	Chronology of Works
李兴钢简介	180	Profile

李兴钢印象

文 / 黄元炤

李兴钢，1969年出生，1991年毕业于天津大学建筑系后，直接进入到中国建筑设计研究院工作至今。2001年任院副总建筑师；他曾参与北京奥运会主会场"鸟巢"的设计与深化工作，并任中方总设计师；2003年在院内成立自己的工作室。李兴钢对建筑中的中国性始终是感兴趣的，所以他关注和思考中国古典园林及其当代性；同时他也关注和运用当代建筑语言——结构与形式、皮层与材料。他企图在这两条主要的路线上，强化个人的设计能量，梳理出自己的价值判断与建筑哲学观。

兴涛接待展示中心，是李兴钢初期建筑实践中一项重要的代表性作品，从这个作品中也隐约看到他内心潜藏的设计追求。他把对中国古典园林的关注与片段的感觉，转化成用墙体去暗示与实现：利用墙体的水平与垂直的连续性转折与延伸，界定出内外空间；以墙体的实，围塑出虚的空间，暗示人的视觉与移动，创造出一个出口或路径，把建筑的流线和空间给带起来，这有点园林中步移景异的意味。一方面突破一般接待中心或枯燥僵化、或过度喧闹的印象，一方面也创造出一个游园的方式。这有别于他在设计兴涛学校时，与更早他在毕业设计里对中国城市和建筑之间的"复合性"关系的思考，所以可以观察到，他对园林的关注和思考，已从这个项目开始，并且也在之后的项目中逐渐体现出来。

"鸟巢"，中方总设计师，是李兴钢给人印象最深刻的一个角色，但在参与"鸟巢"设计工作之前，他曾有一个很好的铺垫，就是西环广场暨西直门交通枢纽的中外合作项目。他当时作为设计的总负责人，主持工程方面的设计和具体工作，在中外合作与城市交通节点的双重复杂因素考量下，有了很好的收获——对大型工程的把握与主持，对技术的控制与执行；同时也建立起一种中外合作的对等关系。一方面中外建筑师相互学习、借鉴与交流，一方面增加中国建筑师的自信心，沟通无障碍，思维国际化。这些方面的经验与提升，成为他参与"鸟巢"设计之前非常好的铺垫。

李兴钢因"鸟巢"的原因，和赫尔佐格与德梅隆接触并且对话。他站在中国建筑师的视角，给予"鸟巢"合作设计团队充分的设计建议与价值参考。当时的李兴钢从多重的角色与角度——诸如建筑师的角度、社会大众的角度、专家的角度、政府和权威者的角度、乃至中国特有的文化角度等等——去判断这个设计是否符合社会的需求与在中国的被接受度，加上他自己多年来职业建筑师的工作经验与设计想象，给予"鸟巢"合作设计团队适当与适宜的参考答案，而这些答案事实证明绝对是有价值、有结果的。当然，在后来的"鸟巢"工程设计和实施阶段，作为中方设计主持人，李兴钢的作用、贡献和所经历的磨炼有目共睹。除此之外，李兴钢也因和赫尔佐格与德梅隆的接触与合作，进而学习到他们的工作方法与工作模式——怎样使一个建筑的设计和实施从开始到最后，逐渐发生、发展、完成。如何把事前工作当成设计线索来准备，如何控制每个设计的工作模式，每一步要做什么，直到最后产生一个出色的建筑。在当时的环境与氛围里，所有的环节李兴钢都参与其中，自然受到的影响很大。之后李兴钢成立自己的工作室，他也建立起这样一种类似的工作方法与工作模式，建立起自己的团队。

Portrait

By Huang Yuanzhao

Li Xinggang, born in 1969, after graduating from Department of Architecture, Tianjin University in 1991, he directly began to work in China Architecture Design & Research Group (CAG) until now. In 2001, he acted as the vice chief architect of CAG. In 2003, in CAG, he established his own atelier. Li Xinggang is always interested in the Chinese feature in architecture, so he focuses on the classical Chinese gardens & their contemporary characteristics, as well as applies the modern architecture languages – structure and form, skin and material. He intends to strengthen personal design power, and express his own value judgment and architectural philosophy on these two main routes.

Xingtao Reception & Exhibition Center, was one of the most important work in his initial practice. From this work, we can seek out his design pursuit hidden in his mind. His attention to the classical Chinese gardens and feeling to segments were implied and realized with walls: the horizontal and vertical continuity and extension of walls were used to define the interior and exterior spaces, indicating the vision and movement of people and combining architecture streamline and space effectively, which seems like scene changed with the position of visitors in Chinese gardens. We can observe that his focus and thinking on the gardens has started from this work and will gradually continue in the works afterwards.

In Xihuan Plaza & Xizhimen Transportation Hub, as the chief designer, he presided engineering design and particular work. Undergoing the complication of Sino-foreign cooperation and urban traffic node, he had harvested much, namely the master and control of large projects, the control and execution of tecnologies. Moreover, by establishing an equal relationship, the mutual learning, reference and exchange among Chinese and foreign architects were available, and the Chinese increased their self-confidences and international thoughts, and could communicate with the foreigners freely. The experience in and improvement from such aspects granted him the very good basis before his participation in the design of "Bird's Nest".

By virtue of "Bird's Nest", he had touched and dialogued with Herzog & de Meuron (HdeM). At that time, from multiple roles and angles, such as the angles of an architect, the public, an expert, the government and authority, and the Chinese-featured culture and so on, he determined whether this design was in line with social demand and acceptable in China. With his work experience and design imagination of many years as a professional architect, he gave the "Bird's Nest" design team with suitable and proper reference, which, proved by facts, were absolutely valuable and fruitful. Of course, at the engineering design and implementation stage of "Bird's Nest" later, as the chief designer of the Chinese party, Li Xinggang's role, contribution and experience were remarkable. Besides, via the touch and cooperation with HdeM, he learned their working methods and mode–how to gradually start, develop and complete a great design and implement of a building from the beginning to the end. In the environment and atmosphere at that time, he participated all steps in the whole process and accordingly he was greatly affected. Later, he established his own atelier and also set up such similar working methods and mode, and his own team.

皮层，是"鸟巢"所体现出来的设计语言。"鸟巢"那由不规则钢结构编织而成的椭圆马鞍形，彷佛是从瓷器的古雅意韵中衍生出的外形，而就建筑学的视点，"鸟巢"表现出来的就是一个"结构性"皮层的建筑语言。李兴钢也因此开始关注到这个极富当代性的建筑语言，皮层的应用，也在他的其他项目中依稀可见。在复兴路乙59-1号改造中，李兴钢设想利用皮层与建筑内部空间产生联系，形成一种透视延伸的感觉，带有"皮层—空间"的设计倾向。而这个内部空间，被他想象成一个垂直方向的园林，有人停留的地方，也有行动的空间，而行动的过程和停留之处的外面皮层是不一样的——它们的透明度不一样，也就是说，皮层跟路径结合，跟视觉感官发生了关联，不同透明是对内部空间的不同暗示；对人来说，则有了不同的景观体验。这种对于结构网格及材料的发现和思考，是李兴钢从赫尔佐格与德梅隆那里学习到的——如何关注材料本身的潜力，对材料进行再开发与设计。在北京地铁4号线出入口中，体现的是"皮层—造型"的设计倾向，在结构网格由小渐大的重复动作中逐渐形成一种造型，演化成多面向与多角度的整体感，组织出新的物质性产物。其实"鸟巢"也是相同的倾向，所表述的完全是一个皮层造型体，建筑彷佛变成是一件艺术品。观察这几个项目，就会发现有一个共同的设计语言：皮层。那是因为李兴钢在参与"鸟巢"设计工作的同时，自己工作室的独立设计项目也在进行，相互之间多少会受到影响。

中国性如何在当代建筑中体现，是李兴钢始终思考的命题。而对于这个中国性，他早期曾关注城市和建筑之间的"复合性"，之后则关注中国古典园林的当代性，如兴涛接待展示中心，创造出一个当代商业建筑中的"游园"方式和空间；如大兴文化中心，建筑围绕着一个立体变化的园林式庭院布局。而在这个过程当中，他又转而关注皮层的建筑语言，但对于园林的思考，始终不曾中断过。所以，他开始想把这两条路线做一个结合，企图从这当中去强化个人的创作能量，梳理出一条明确、清晰的设计路线。复兴路乙59-1号改造，是这两条路线结合的初探：皮层与垂直园林的结合，但皮层语言表现太过强烈，锁住了人们的视觉。在建川镜鉴博物馆中，这两条路线的结合更为明显，在立面上用砖来做皮层的处理，塑造出不同通透感的砖砌墙面与内部功能进行对应和暗示，同时用砖来达到一种封闭性与内向性，以突出内部庭院的景观和体验；而内部空间中参观展览的过程则是园林式的体验，运用园林的元素，直接或抽象地表述出来，并使用"复廊"和"亭台"的手法，构成线性的展览空间，所以实际上还是成为了一个繁复的游园式空间。而镜子，在这个项目中是最好的媒介，它回应于建筑的主题，也被设计成装置化与物质化的表现物件——"镜门"，它制造虚幻的氛围，让人亲身体会一种吸引与强迫的历史场景。

综观李兴钢的设计，似乎摆荡在现代与传统之间，或者是企图融合两者。传统，绝对是他所关注的园林，并从中领悟与寻找当代性的表达；而他所认知的现代，更靠近了当代与未来——一种当代的皮层，结构与形式的相互激发与转化，材料潜力的挖掘与再创造。总之，李兴钢逐渐在形塑他自己的建筑语言系统或者是哲学观，他从原本放任的姿态，到慢慢有意识且主动地去思考与建立。步伐，循序渐进与不疾不徐；态度，悠游放松与随性自在。他觉得这样建立的过程是自然而然的，是经验积累的，结果也应该是顺从人意的。下一步，就来看看李兴钢如何激发自己的设计能量吧。

Skin is the design language expressed by "Bird's Nest". From the point of architecture, what "Bird's Nest" expresses is a "structural" skin. Thereby, Li Xinggang starts to pay attention to this modern architectural language, the operation of skin, which glimmers in his other projects. In Reconstruction of No. B-59-1, Fuxing Road, he conceived connecting the skin and the interior space so as to form a feeling of perspective extension with the design inclination of "skin-space". This interior space was regarded by him as a garden in vertical direction; the place with the stay and the space of action corresponded to different exterior skin – so were their transparencies. Namely, the combination of skin with path was related to organs and sense of vision, and different transparency was the different indication to the interior space. This discovery and thinking of structural grids and materials were learnt from HdeM, i.e. how to note the potential of materials and re-development & re-design of them. Accesses of Line 4 of Beijing Subway expresses the design inclination of "skin-shape". In the repeated action of structural grids from smaller to bigger, a shape gradually comes into being, evolving into the multi-directional and multi-angular integrity. The expression is completely a skin-shape body and seems an artwork. Observing such several projects, we can find the same design language: skin. While he was participating in the design of "Bird's Nest", the independent design projects of his atelier were ongoing. As a result, they were influenced by each other more or less.

How to indicate the Chinese feature in the contemporary architecture is the subject Li Xinggang is always thinking about. In respect of this Chinese feature, in the early time, he paid attention to the "compound" between city and architecture. Later, he shifted his attention to the contemporary characteristics of the classical Chinese gardens. For example, Xingtao Reception & Exhibition Center has created one "garden-sightseeing" mode and space in a modern commercial building. In such course, he again shifted his attention to the architectural language of skin. However, he has never interrupted his thinking about Chinese gardens. As a result, he started to combine these two routes in order to reinforce personal creation power therein and obtain a clear and remarkable design route. Reconstruction of No.B-59-1, Fuxing Road was the primary test of the combination of these two routes, i.e. the combination of skin and vertical garden; however, the skin language was too remarkable, locking the vision of people. In Jianchuan the Cultural Revolution Museum & Wenchuan Earthquake Memorial, the combination of these two routes was even more remarkable. Bricks were used to treat the skin on the facade, creating the leaky brick-wall surface of different transparencies, corresponding and indicating the interior functions. Bricks were also used to realize the effect of enclosure and introversion in order to stress the scene and experience. With the method of "double-corridor" and "pavilion", the linear exhibition space was formed a complicated garden-type space. The mirror, as the best medium in this project, reflected the theme of the building and also was designed as the installation, the "mirror-door", which created a visional atmosphere and allowed people to personally experience an attractive and forced historical scene.

In a whole, the design of Li Xinggang seems swinging between modernism and tradition, or the combination of both. Tradition absolutely is the garden within his attention, from which he feels and looks for the expression of contemporary character. The mondernism as he knows is closer to the contemporary and the future, i.e. a modern skin, the mutual inspiration and conversion between structure and form, and the exploration and recreativity of material potential. In a word, he is gradually constructing his own architectural language system or philosophy. From the original indulgement, he slowly becomes rational and actively thinks & constructs his own system. He takes the orderly and reasonable steps, and holds the attitude of ease and freedom. He thinks this course is natural and needs experience, and the result is satisfactory. Next, let's see how he inspires his own design power.

建筑的发现与呈现

文 / 李兴钢

The Discovery and Presentation of Architecture

By Li Xinggang

对我而言，建筑的神秘在于它早已存在那里，按照使用者的自然天性和建筑自身的朴素逻辑。而所谓设计只不过是在分析了种种给定的条件和多样的可能性后，寻找到那几乎唯一完美的答案。当然，寻找的过程和表达的方式自然带有因人而异的倾向或痕迹，比如，我是一个如此这般的中国人。

在学习和从事建筑18年后，我写下以上的心得。2004年天津大学建筑学院邀我做一次讲座，主题叫做"发现建筑"，我以此作为毕业后向母校和老师的工作报告。我的讲座是由四川三星堆遗址博物馆收藏的四件商代玉器开始的：第一件玉璞、第二件玉戚形璧、第三件玉戚形珮、第四件环形玉璧。如果把第一件璞玉看成这四件玉器的"前身"，可以想象这样的故事和情景：第四个工匠将璞玉雕琢成一个完美的环形，并将玉石的自然纹理呈现于玉璧表面；第三个工匠顺应璞玉原有的形状和肌理进行适度的加工，得到一件呈盾形对称的玉珮；第二个工匠则基本上把璞玉的自然形状和质地肌理完全保留，只在中间做了一个非常精致的圆孔，这是一件非同寻常的玉璧，天然的朴拙与人工的创造如此完美地结合为一体；而第一个工匠觉得璞玉本身已很完美，干脆不施身手，完全保留，几千年之后竟也成为一代绝品。

这是一个发现与呈现的过程：研究现状、发现线索、制定策略、表达呈现。这也是一个由理性到感性的过程：以理性思考开始，渐以感性表达终结。这完全是一个设计过程的描述，今天、今天以前和今天以后，这样的工作方式，被很多建筑师、很多成功的建筑师所运用和继续运用，也应该是不会有错的。但问题是为什么有人成就为工匠中的大师，有人则永远只是平庸的工匠呢？究竟哪件玉器才是那"唯一完美的答案"？实际上就如同一块璞玉只有一次成为某件玉器的机会一样，一个建筑也只有一次机会成为它自己，作为建筑师，你无法从头再来，这是这个职业的遗憾和挑战，也是这个职业的魅力所在。

建筑的发现非常重要，将设计者引向正确的方向，是理想答案产生的必要前提；建筑的呈现更是如此重要，那是决定建筑命运的时刻：平庸之作还是传世精品、使人无动于衷还是心灵激荡？从发现到呈现，往往经历艰苦卓绝的研磨过程，犹如精美玉品的生成。而建筑的最终呈现，则不仅止于设计，还得经由艰苦漫长的建造而矗立于大地和城市，并最终为人使用和检验，方始完成。

建筑的本质是什么？好建筑是否存在恒久的标准？我因此类问题而经常周期性地处于迷惑—清晰—迷惑的状态，但至少以下几点于我是清晰的：风格不是建筑的（而常常是商业的）目的；建筑的设计应该由回答之所以要存在开始，并回应人类的本能；好建筑会在斯时、斯地、斯境与体验者或使用者达至心灵的契合；好的职业建筑师身处自己的时代和特定的社会、文化背景，应该以自己的思考和实践推动建筑学（或其某一方面）的发展，并以此尽到对于人类生活的责任。

一次在工地的偶然经历，曾使我惊诧以至沉思良久：那是一组民工们临时搭建的工棚，却具备很多好建筑的元素和品质：形式、空间、材料、构造、

To me, the enigma of Architecture exists in its longevity and antiquation – its endurance and history are the result of natural instinct and plain logic. The so-called process of "design" is the journey to find the almost one and only perfect solution, solely realized after analyzing the various given conditions and possibilities. Naturally, the journey and destination bear the tendency or impressions of the architect, in this case me – a specific Chinese person.

After 18 years of study and working on architecture, I gave a report named "Discovering Architecture" to my Alma Mater in 2004. The lecture began with four jade articles of the Shang Dynasty within the collection of the Sanxingdui Ruins Museum of Sichuan Province. They include a Yupu (Rough Jade), a Jade Bi in the Shape of a Qi Axe, a Jade Pendant in the Shape of a Qi Axe, and a Ring-shaped Jade Bi. If the rough jade is deemed as the "precursor" of the other three articles, the following story can be imagined: the forth craftsman carved the rough jade into a perfect ring and presented the natural texture of jade on the surface; the third craftsman appropriately processed the rough jade according to its original shape and texture and got a shield-shaped and symmetrical Jade Pendant; and the second craftsman generally kept the natural shape, and texture of the rough jade and only drilled a delicate hole, which perfectly combines the natural roughness with artificial creativity; but, the first craftsman, thinking the rough jade was inherently perfect, purely did nothing at all on the the Rough Jade, which, after thousands of years, unexpectedly became a nonesuch.

This is a process of discovery and presentation: studying the present situation, discovering clues, customizing strategy, and presentation. It is also a process starting from rational thinking to emotional expression. Such description of the design process, although plain, should be true because it was, is, and will be used by many architects, including successful ones. But why have some craftsmen become masters but others are always mediocre? What on earth is the "one and only perfect solution"? Just as a rough jade only has one chance to become a jade ware, a building only has one chance to be itself. As an architect, you can never start over again. This is the drawback and challenge but also the charm of this profession.

The architectural discovery is important because it leads the direction and is the prerequisite to generate an ideal solution. The architectural presentation is even more important because it determines the destiny of architecture and whether it is mediocre or excellent. The process from discovery to presentation is often the one as hard as the generation of an elegant jade ware. Moreover, the final presentation of architecture does not end at the design. It cannot be completed until the building has been erected used and verified by people finally.

What is the essence of architecture? Is there any permanent criteria of good architecture? These questions periodically puzzle me, although I clearly understand the following points: style is not the purpose of architecture (but of business); architectural design should first answer why a building should exist and should respond to human instinct; a good building should be harmonious with the soul of users at a given time and place and in a given context; and a good professional architect, who is within his/her time, social and cultural background, should assume his/her responsibility by promoting the development of architectonics (or some aspect of it) with his/her thinking and practice.

细节乃至建筑的神态和气质。简单而内敛大气，放松而自然到位，犹如大师之作，它使周围那些真正要建造的建筑师"作品"（也包括我的）相形见绌。我跟刘家琨建筑师聊天时讲到此，他把这样的情况叫做"素人建筑"。

阿尔瓦罗·西扎在一篇文章开头写道："毕加索说他花了十年时间学会了绘画，又花了另外十年学会了像孩童般的画画。现今，在建筑学的训练中缺少了这后一个十年。"在我们长大成人接受各种教育的过程中，其实也丧失掉很多天性和创造性，难得的是能够最大程度地保持自己，并在创作活动中体现符合人的本能和事物本原状态的创造性，再加入独特的艺术判断力，才会使作品格外有力量，给人以震撼。伟大的建筑师路易斯·康说，"我爱起点，一切人类活动的起点是其最为动人的时刻"。我认为，并不止于建筑，这也许是值得所有艺术回头深思并有所启示的地方。扪心自问：我们身在何处？我们已经离我们自己有多远？

建筑学成为一个高高在上的"专业"，我以为是一件可悲的事情，这当然是另外的话题。对一个好建筑师来说，也许最后他会发现，所谓的专业技能对他来说并非是最重要的东西，一个建筑师首先应该成为一个真正的人，一个情感丰富、博学通达、敏感灵慧的人，一个对身边事物和人性能够深切感知、体察的人，一个对文化、社会怀有批判精神和责任感的人，一个始终具备、发展自己的创造力和艺术直觉的人，一个具有敬业精神和工作热情的人，当然，还应该是一个健康、平和、享受生命快乐的人。文如其人，建筑也如其人，因此，我越来越感到自己这个人对于建筑设计的重要，做人的质量决定着做建筑的质量——归根到底，什么样的人，就会做出什么样的建筑。设计的活力来自生命的活力，生命的活力来自对生活的敏锐感知和永不休止的思想。

与我们的生活和建筑的呈现相关的一个话题是文化和传统。非物体之美、空灵之美、平和之美、意境之美，跟这些美相关的人的生活和生活方式，是我们失掉和正在失掉的可贵传统，是我们东方人的传统。我不是一个很激进的人，否则的话我就会说，也许这种美和这种东方的哲学，是挽救我们的城市、挽救我们这些生活在城市中的人、甚至拔高一点讲，是挽救我们文化的唯一的出路。作为生活在这个地方的人，也许东方文化是最适合我们的，生活在一个全球化的、浮躁的现代商业社会里，可能我们并不能够意识到这一点，但是它是在我们心里的，是在我们潜意识里的。

对园林和聚落的体验和研究，给我们很多的感触和启示：从中可以看到和感到令人惊讶的当代性，中国特有的文化魅力和生活哲学活生生地搁在那里，在这个文化浮躁、混乱、茫然的当代社会，在那里我们会找到自己的文化之根和生活自信。我们企图发现其中那超越时间、令古人与今人通感心灵激荡的秘密，相信会对我们的设计有不同寻常的启发。最终，在当下复杂纷纭、极具挑战性的中国现实中，在建筑中发现和呈现我们的文化美学、判断力和自信心，这是我们的使命和职业理想。

我把下面四句话写给自己和伙伴们，并与所有从事这个职业的人们共勉：建筑既是不断研磨的发现，又是不可言说的呈现；建筑的工作既带给你快乐，又无法避免的长时间的繁冗琐碎；建筑不是高高在上的专业，而是建筑师作为一个人的真实自然的表达；建筑不是纸上谈兵，而是亲身的建造、体验、触摸和攀爬。

I was surprised and made thinking for a long time by an unexpected experience on a construction site. That was a group of temporary work sheds built by constructors. These sheds had many elements and characters of good architecture in terms of the shape, space, materials, construction, details, and even architectural expression and temperament. They were unsophisticated and self-restrained but have nothing petty. They were also relaxing and naturally in place. Just like the work of a master, it was superior over the desired "works" of architects (including mine) that were to be built nearby. When I mentioned them to architect Liu Jiakun, he called them "natural people's architecture".

"Picasso said he took ten years to learn to draw, and then ten more to learn to draw like a child, the architectural training lacks of this second ten years nowadays", said Alvaro Siza. When we grow up and are educated, we in fact lose much of our nature and creativity. It is rare to keep ourselves to the greatest extent and represent the creativity that fits in with human instinct and the original state of things and the unique artistic judgment is attached, which makes our works powerful and impressive. The great architect Louis I. Kahn said that he loved the origin because it was the most engaging time of all human activities. I believe that this is where architects and all other artists should rethink and get an idea. We should answer the questions: Where are we? How far have we been away from ourselves?

In my opinion, it is no good for architecture to become a profession up above. A good architect may finally find that the so-called professional skills is less important for him/her; instead, he/she should first become a real person who has emotions, extensive knowledge & understandings, sensibility, and intelligence; who is critical to and responsible for culture and society; who has and continuously develops his/her creativity and artistic intuition; who is dedicated and enthusiastic to his/her job; and who is healthy, peaceful, and enjoying life as well. Just like an article mirrors its writer, a building shows its architect. This is why I increasingly feel that the quality of me as a person decides the quality of the buildings. The activity of design comes from the activity of life, which originates from the sensitivity and endless thinking of life.

One of the topics related to our life and architectural presentation is culture and tradition. The life and lifestyle related to the beauty of intangibility, spaciousness, peacefulness, and conception are the invaluable tradition that our easterners have lost or are losing. If I were radical, I would say that the beauty and eastern philosophy may be the only way to save our cities, save us living in cities, or even our culture. The eastern culture may be most suitable to us who live here, although we may not realize it because we are living in a global, fickle, and commercial society. However, it is in our heart and subconscious.

In fact, we get much feel and inspiration from gardens and settlements: we can see and feel the surprising contemporary, where the unique Chinese culture and living philosophy exist. They are the place where we can find the root of our culture and the confidence to our life in today's society that is full of fickle, confused, and vague culture. We attempt to find the secret that makes gardens and settlements going beyond time and reaching resonance between people of the past and present. We believe that this will provide extraordinary hint to our design. In the complex and challenge reality society in China, it will be our mission and ideal to finally discover and present our cultural aesthetics, judgment, and self-confidence in architecture.

The following statements encourage me and associates in career: Architecture is continuous discovery and unspeakable presentation; Architecture is an enjoyable job despite the long-time redundant and trivial efforts; Architecture is not a profession up above, but the natural expression made by an architect as a human; and Architecture is nothing of empty talk or drawing on paper, but practical construction, experience, touch, and scrambling.

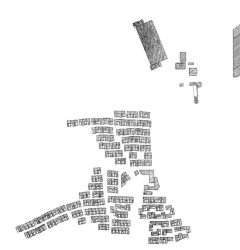

由入口看小区大门 / View of gate from entrance

兴涛接待展示中心　　北京

Xingtao Reception & Exhibition Center, Beijing
2001

兴涛接待展示中心位于北京郊区一个正在开发中的商品住宅小区的入口处。包含了接待、展示、洽谈、住宅样板间、小区大门及警卫室等功能。

设计将整个建筑分为两组体量，分置在狭长的用地两端，中间以长墙／廊相连，由于展示、接待、参观、洽谈和销售的流程而产生了一种动态的流线，来实现建筑的使用过程，这一流线是由墙这一要素的延伸变化来引导的。墙对空间体验的这种动态的引导性是中国建筑传统特别是古典园林的重要特点之一。在中国园林中，由于连续的墙体所特有的导向性，使身处其中的人不由自主地产生一探究竟的欲望，由此在人的运动中发生丰富的空间体验，使中国园林成了真正的四维建筑。在这里，一片白墙由建筑的入口开始，不断地在水平垂直方向延伸运动，忽而为垂直的墙，忽而是水平的板，或升或降，如此形成了建筑的骨架和内外空间；在这个建筑和空间骨架中再插入透明的玻璃体和玻璃廊、灰砖的样板间单元和警卫室以及一片黑色的浅水池。景观（水面／庭园）设计与功能流线和空间体验有着密切的关系。

这个小建筑试图将它特有的商业特征与中国传统园林的空间体验和东方意味融合在一起，用一种有趣的、传统的方式来实现商业的、现代的功能，并使用当地的、现时／现代的、可操作的技术满足业主的现实需求和低造价下的快速建造。

摄影：张广源

Xingtao Reception & Exhibition Center is located at the entrance to residential quarter in suburban Beijing. It provides reception, exhibition, and negotiation functions, a residential sample unit, and the gate & guardroom of the quarter.

The entire building is divided into two masses at each end of the parcel but is joined together by a long wall/corridor, which extends and changes to guide a dynamic flow of exhibition, reception, visiting, negotiation, and sale. Such dynamic guidance of walls to the spatial experience is one of the important characteristics of Chinese traditional architecture where continuous walls are unique guides that inspire visitors to find all things out involuntarily. When they move, the visitors gain abundant experience of the space and the gardens become truly four-dimensional buildings. This is just the case in Xingtao Center, where a white wall starting from the entrance continuously extends and moves in the horizontal and vertical direction. Although this vertical wall changes to a horizontal platform at some places, it keeps rising up and down to form the building skeleton and internal & external space in which a transparent glass body and corridor, a grey-brick sample unit and guardroom, and a black shallow pool are inserted.

This building tries to integrate its unique commercial characteristics with the spatial experience and eastern meanings of traditional Chinese gardens so that its commercial and modern functions can be realized in an interesting and traditional manner. Locally available, current/contemporary and operational technologies are used to meet the practical needs of owners and to quickly construct the building at a low cost.

Photographs: Zhang Guangyuan

池边入口 / Entrance by pool

全景 / Panoramic view

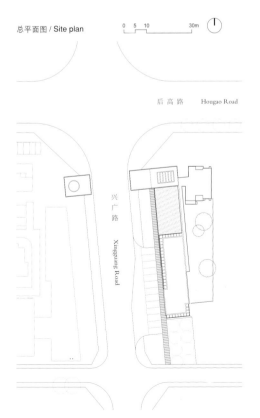

总平面图 / Site plan

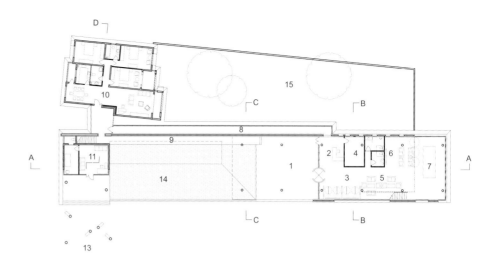

首层平面图 / The 1st floor plan

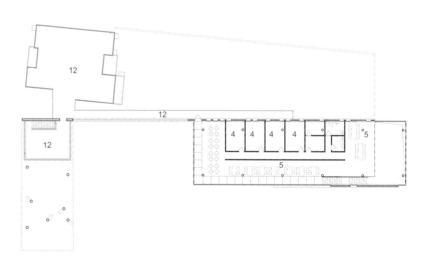

1	入口	Entrance
2	服务台	Reception
3	图纸展板	Exhibition of drawings
4	办公	Office
5	洽谈签约区	Meeting area
6	放映区	Video show
7	模型展示	Exhibition of models
8	玻璃廊	Grass corridor
9	临水平台	Terrace on water
10	样板间	Simple unit
11	警卫室	Guard
12	屋顶平台	Roof terrace
13	小区大门	Gate of quarter
14	水池	Pool
15	庭园	Courtyard

二层平面图 / The 2nd floor plan

15

夜景 / Night view

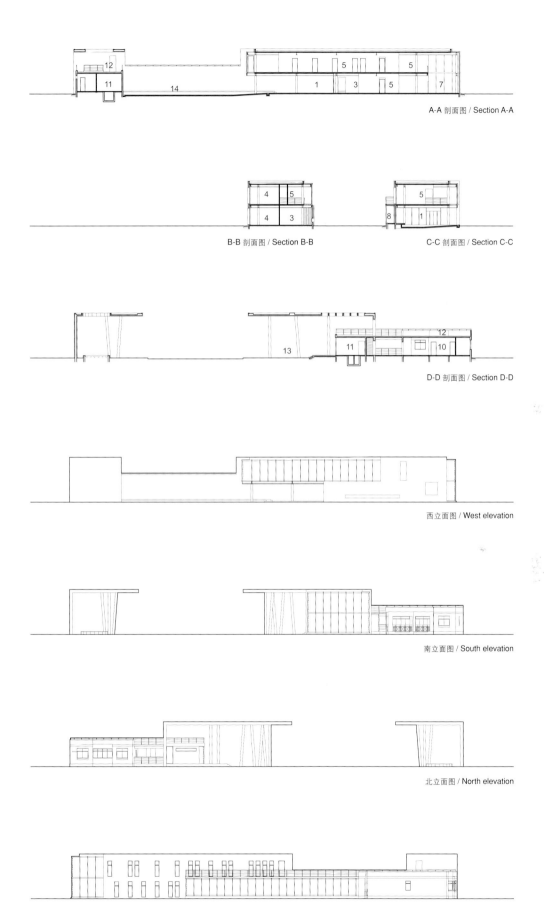

由小区大门看接待中心 / View of reception center from gate

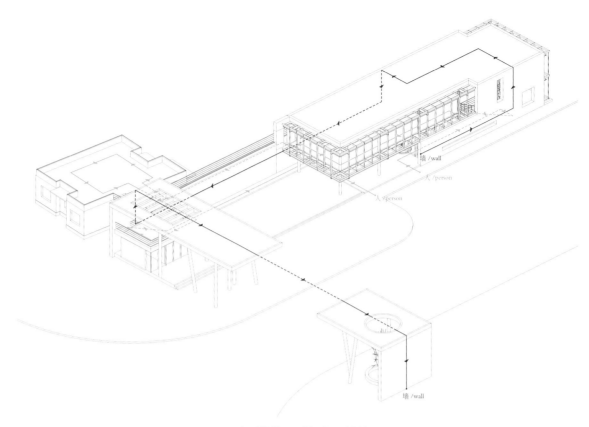

"运动的墙板 & 运动的人"——轴测分析图 / "Moving Wall & Moving Person" axonometrical analysis

小区大门 / Gate of quarter

玻璃廊 / Glass corridor

街坊中间的步行小巷 / Alley between blocks

建川镜鉴博物馆暨汶川地震纪念馆

安仁

Jianchuan Mirror Museum & Wenchuan Earthquake Memorial, Anren
2004～2010

建川镜鉴博物馆暨汶川地震纪念馆位于四川大邑安仁古镇,是民间投资建设的建川博物馆聚落中的单体建筑之一,最初设计为镜鉴博物馆,收藏展示文革时期的各种镜面。汶川地震发生后,通过设计改造增加了地震纪念馆,以"震撼日记"的形式收藏和展示地震文物和相关艺术创作,成为一个两馆一体的复合型博物馆。

原设计建造并停工难产的文革镜鉴馆由于汶川地震的发生加入了地震纪念馆而再造重生,经设计改造后的两个馆在空间上相互叠加并置,"虚像"和"现实"相互混合、对照。镜鉴馆通过组合的镜门装置造成纯净、抽象、变幻多端的虚像空间而让参观者以游戏的方式模拟体验失去理智的疯狂年代;地震馆则以临时、粗粝、具体、真实的空间和展品让参观者感受痛切而震撼的现实。两者以各自的方式纪念、展示和体验着历史上发生的"人祸"和"天灾"两大人间悲剧,给予后人以鉴戒和警示。建筑外部平和而宁静,却混合收藏着内部的虚像狂乱和现实震荡,给予当代人对过往历史和灾难的即时体验与纪念,加上其曲折、多变、富于现实感和戏剧性的设计建造历程,可称得上是这个时代的"镜鉴"之馆。

设计使用的主要外墙材料是清水混凝土和红、青两色的页岩砖,当地丰富的砌砖传统在这里体现为在同一砖砌模数单元控制下,不同的室内功能对应不同通透程度的砖砌花墙,以满足不同的采光、通风、景观和私密性等要求。为此还专门设计发明了符合上述模数的透明"钢板玻璃砖",造价低廉并易于加工,用于花墙上对应室内空间的砌空部分。

摄影:张广源、李兴钢

Jianchuan Mirror Museum & Wenchuan Earthquake Memorial is located at the ancient town of Anren in Dayi County, Sichuan Province. It is one of the single buildings among the non-governmentally funded Jianchuan museum cluster. The museum was originally designed to collect and exhibit mirrors from the Cultural Revolution of China. After the Wenchuan Earthquake, it was redesigned into a composite museum in which Wenchuan Earthquake relics and the relevant artworks are exhibited in the form of a "Shocking Diary".

With a modified design, the museum and memorial are spatially overlapped and the "virtual images" are mixed and in contrast to the "reality". The Mirror Museum creates a pure, abstract, and constantly changing space using a composite device of mirror-doors. The space consists of virtual images that imply the crazy years of Cultural Revolution for visitors to gain a simulated experience. However, the Earthquake Memorial lets visitors sense the grievous and shocking reality with temporary, tough, concrete, and real space & exhibits. The museum and memorial show, and enable visitors to experience the two disasters – the manmade Cultural Revolution and the natural Wenchuan Earthquake – in different manners, thereby providing later generations with warning and alert. The interior mixture of crazy virtual images and shocking reality are contained in the peaceful and tranquil exterior, which allows people to instantly experience the history and disasters.

The exterior wall is mainly designed with as-cast finish concrete and shale bricks of red and grey. Leaky brick-walls of the same modular bricklaying unit embody the rich brickwork tradition of the local place. Their degree of opening varies with the lighting, ventilation, landscape, and privacy needs of different indoor functions. Cheap, easy-to-process, and transparent "steel plate + glass bricks" meeting the above module have been designed and invented for leaky brick-walls openings corresponding to the interior space.

Photographs: Zhang Guangyuan, Li Xinggang

西立面 / West facade

总平面图 / Site plan

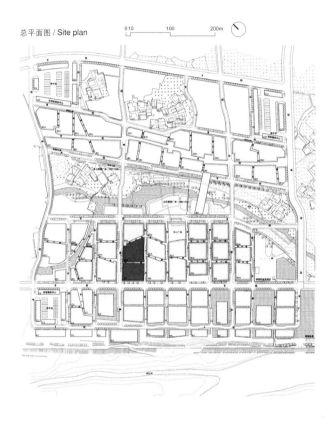

模型研究 / Model study

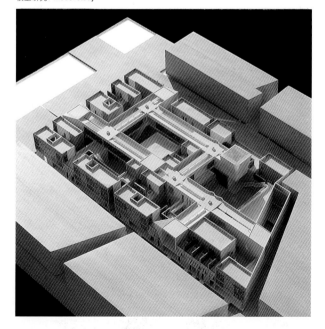

西北俯瞰 / Northwest aerial view

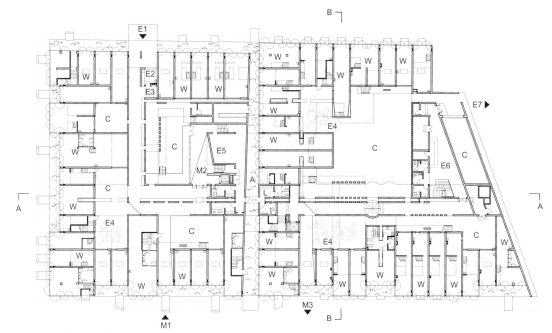

首层平面图 / The 1st floor plan

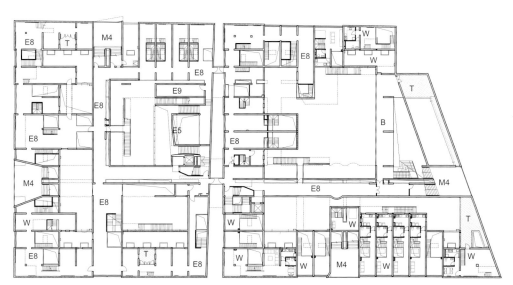

夹层平面图 / Mezzanine plan

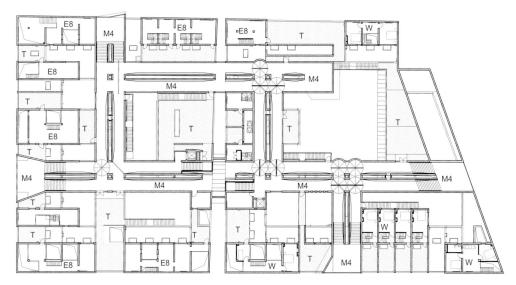

二层平面图 / The 2nd floor plan

M1	镜鉴馆入口	Mirror Museum entrance
M2	序言厅	Preface hall
M3	镜鉴馆出口	Mirror Museum exit
M4	镜鉴馆展厅	Mirror Museum exhibition hall
E1	地震馆入口	Earthquake Memorial entrance
E2	纪念品商店	Memento shop
E3	售票处	Ticket office
E4	室外展场	Outdoor exhibition
E5	罹难者照片墙	Wall of victims' photoes
E6	结束厅	Ending hall
E7	地震馆出口	Earthquake Memorial exit
E8	地震馆展厅	Earthquake Memorial exhibition hall
E9	地震体验厅	Earthquake experience hall
W	艺术家工作室	Artist workshop
A	街巷	Alley
C	庭院	Courtyard
L	公共卫生间	Lavatory for public
T	屋顶平台	Terrace
B	茶室	Beverage shop

关闭的铝板隔扇门和支摘窗 / Closed aluminium doors & windows

主庭院 / Main courtyard

A-A 剖面图 / Section A-A

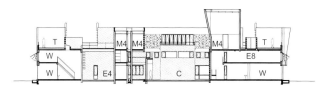

B-B 剖面图 / Section B-B

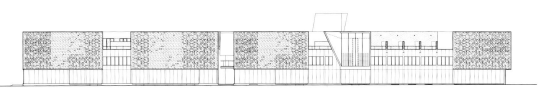

东立面图 / East elevation

0 1　5　　10m

南立面图 / South elevation

镜门与复廊模型研究 / Model study of mirror-door & double-corridor

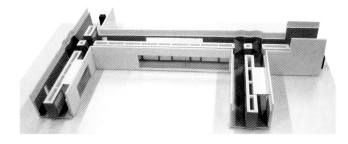

草图示意最佳参观路径 / Sketch of one-way visiting route

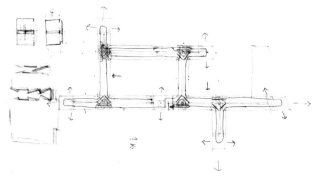

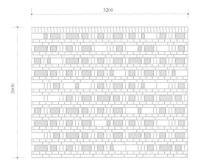

花墙 / Pattern of leaky brick-wall

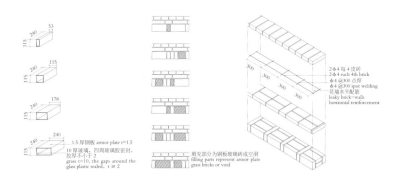

砖及钢板玻璃砖 / Bricks & "steel plate + glass bricks"

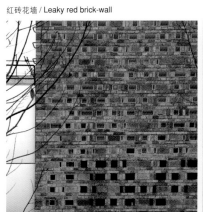

红砖花墙 / Leaky red brick-wall

青砖花墙 / Leaky grey brick-wall

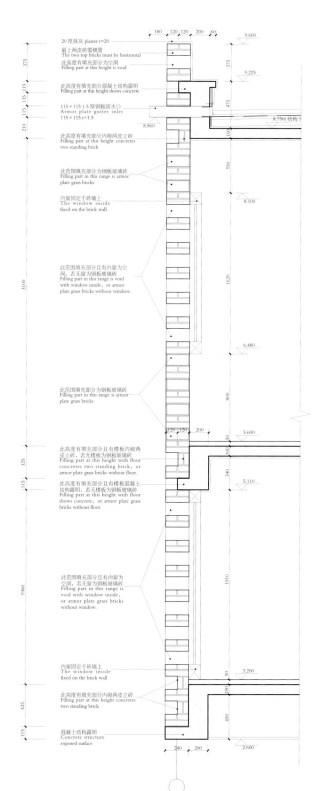

花墙详图 / Details of leaky brick-wall

庭院 / Courtyard

由镜鉴馆展厅看主庭院 / View of main courtyard from exhibition hall of Mirror Museum

利用夹层藏品库改造成的地震馆"震撼日记"展厅 / "Shocking Diary" exhibition hall of Earthquake Memorial reconstructed from mezzanine storage

镜门 / Mirror-door

精品展厅 / Special exhibition hall

立面局部 / Part of facade

复兴路乙 59-1 号改造　　北京

Reconstruction of No. B-59-1, Fuxing Road, Beijing
2004～2007

复兴路乙 59-1 号位于北京长安街西延长线复兴路北侧，原是一幢 20 世纪 90 年代初期设计建造的、9 层钢筋混凝土框架结构的办公和公寓建筑，并在东侧与一栋 9 层住宅楼相连。业主希望在保持原有建筑高度、结构、设施基本不变的基础上，对功能、空间和外观进行整理改善，将其改造为集餐饮、办公、展廊为一体的小型城市复合体。

改造后的建筑体形基于原来的方形体量，并结合周边环境和日照关系进行局部的切削和增长；建筑外幕墙框架网格的生成基于原有建筑较无规律的立体结构体系（不规则的层高分布和变化多样的柱距），并用来作为立面及内部空间的控制系统，既符合改造加建的结构逻辑，天然地反映着原建筑的基本状况，又形成有自身独立特征的结构和形象语言。将生成的外部幕墙网格立体化和空间化，在不同朝向形成不同进深和特征的内部空间，以配合不同的使用和景观要求。西侧利用原室外疏散梯扩展改造成为一个立体画廊，不同高度、位置、形态和景观的展厅和平台被多样的楼梯、踏步和台阶联系起来，由下至上可一直延伸到局部加建的顶层及屋顶庭院，可被视为一个垂直方向游赏的小型园林。

对应内部功能和空间的不同，选用了具有四种不同透明度的白色彩釉玻璃，作为覆盖在外部网格上的幕墙材料，既控制着光线在建筑内外的投射和透射，也左右着人的视线在建筑内外的驻留和延伸。不均匀的透明度加上全隐框的白色彩釉玻璃幕墙，使得建筑呈现出深邃、平静而丰富的气质。

No.B-59-1, Fuxing Road is situated to the north of Fuxing Road, the western extension of Chang'an Street, Beijing. It was originally a 9-storey office and apartment building of reinforced concrete frame construction in 1990s, and a 9-storey residence to the east. The owner hoped to turn the building into a small urban complex that provides restaurant, offices, and gallery by coordinating and improving its function, space, and appearance without much changing its height, structure, and facilities.

The reconstructed building is based on the former square mass but partially modified to coordinate to the surroundings and the sunlight relationship. Frame grids of the exterior curtain wall are generated according to the irregular stereoscopic structural system of the former building. The grids are also used as the control system of the facade and interior space. They not only comply with the former structural logic, but also form structural and visual language with their independent characteristics. The grids create interior stereoscopic spaces of different depth and characteristics to meet different function and landscape. On the west side, the former outdoor emergency stair is expanded and modified into a stereoscopic gallery, which connects exhibition halls and platforms of different heights, shapes, and sceneries. Extending from the ground floor to the locally-added top floor and roof yard on the top, the stereoscopic gallery can be seen as a small garden for tour in the vertical direction.

Corresponding to the different interior functions four types of white glazed glass of different transparencies have been selected as the curtain wall material. The glass controls the casting and transmission of light and dominates the stay and extension of sight in and out of the building. The white glazed glass curtain wall of uneven transparency and fully invisible frame provides the building with a deep, tranquil, and varied temperament.

摄影：张广源、李兴钢、付邦保

Photographs: Zhang Guangyuan, Li Xinggang, Fu Bangbao

不同透明度的幕墙玻璃 / Curtain wall glasses with different transparence

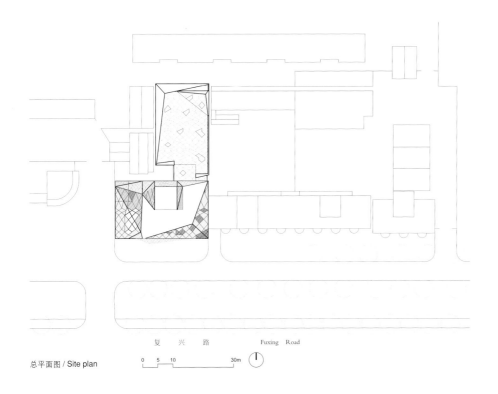

总平面图 / Site plan

复兴路　Fuxing Road

0 5 10   30m

模型研究 / Model study

街景 / View from main street 改造前建筑 / Building before reconstruction

夜景 / Night view

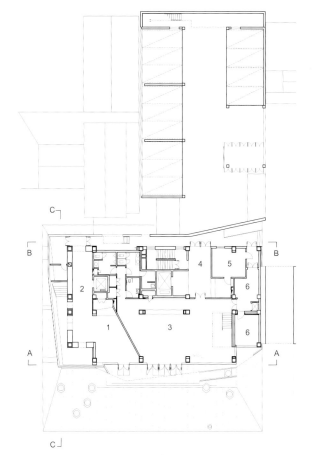

首层平面图 / The 1st floor plan

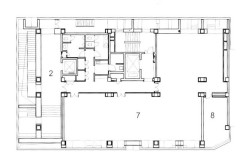

标准层平面图 / Standard floor plan

屋顶平面图 / Roof plan

1	画廊门厅	Lobby of gallery
2	立体画廊	Stereoscopic gallery
3	咖啡厅	Cafe
4	办公门厅	Lobby of office
5	消防控制室	Fireproof control room
6	预留西餐厨房	Reserved for western kitchen
7	办公区	Office
8	休息区	Break area
9	屋顶庭院	Roof yard

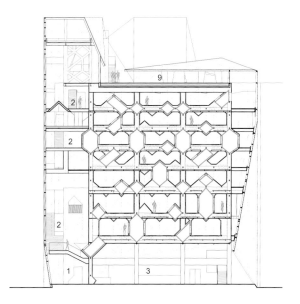

A-A 剖面图 / Section A-A

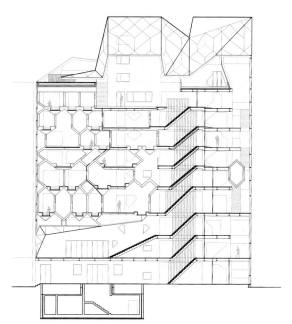

B-B 剖面图 / Section B-B

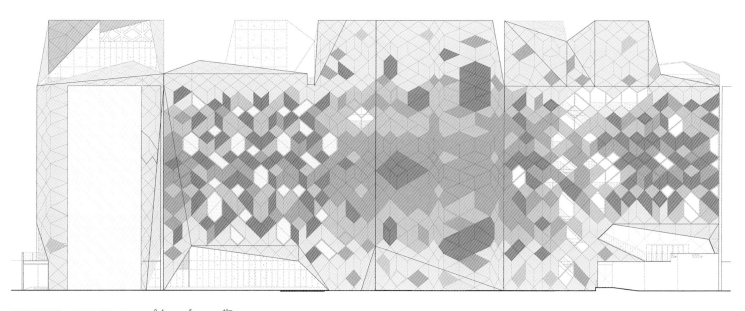

立面展开图 / Spread-out of facades

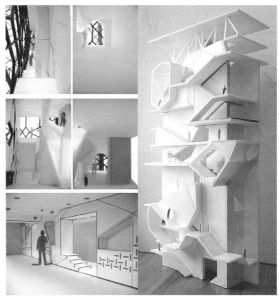

立体画廊模型研究 / Model study of stereoscopic gallery

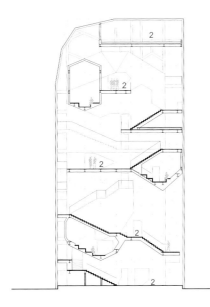

C-C 剖面图 / Section C-C

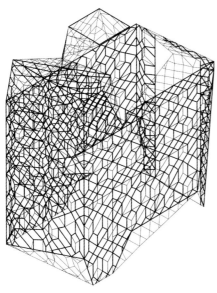

幕墙框架网格 / Frames of curtain wall

立体画廊内景 / Interior of stereoscopic gallery

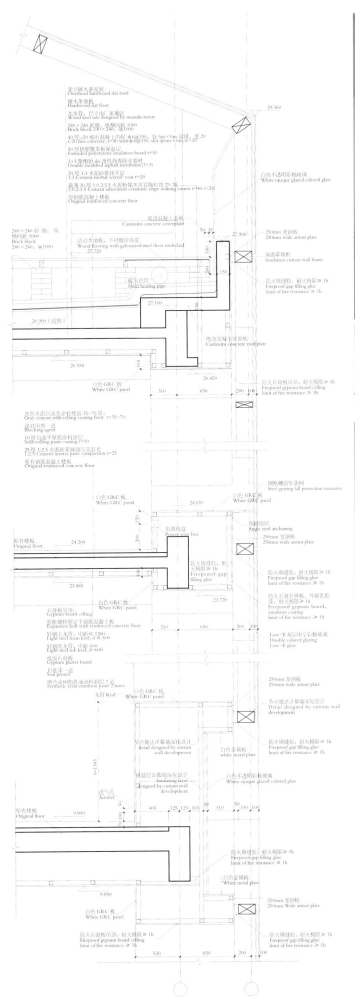

墙身详图 / Details of wall
立体画廊 / Stereoscopic gallery

屋顶庭院 / Roof yard

南立面局部 / Detail of south facade

画廊顶层局部 / Detail of top floor of gallery

立体画廊施工 / Construction of stereoscopic gallery

幕墙实体模型 / Mock-up of curtain wall

西立面 / West facade

Hiland·名座 威海
Hiland·Mingzuo, Weihai
2006 ~ 2010

　　Hiland·名座位于山东半岛威海市城市干道海滨路和渔港路的交口处，邻近东部海滨，是一座以SOHO办公为主、兼具商业功能的房地产开发建筑。

　　设计根据当地的主导风向（夏季以东南风南风为主，冬季以西北风为主），利用气流的基本原理（对流、气压差），以低技的、简单直接的自然通风方式，在建筑内设定了多个"西南－东北"走势的"风径"，从而有效引入夏季风穿过建筑内部以降温除湿，同时最大限度地回避冬季风对建筑的不利影响，并在设计过程中使用了CFD计算机模拟技术对其风速、温度、湿度等进行舒适度模拟验证和校核，使建筑内尽量多的房间能够在夏季通过自然通风的方式降温，从而减少空调设备在夏季的使用。"风径"口部设置了可密闭的旋转门，根据外部天气的变化调整门的开闭情况以产生舒适的建筑小气候，实现建筑室内外空间的自然转换。"风径"的设置不仅可以有效地改善建筑内部小气候——降温除湿，也形成了积极的邻里交往空间和极佳的观景平台，使自然环境和人的活动在建筑中交融共存，也使建筑以一种开放的姿态和形象存在。

　　建筑同时注重与城市环境的对话，红色坡屋顶的设计不仅呼应了威海市特有的红瓦坡顶建筑风格，同时也有效减少了建筑对基地南侧、东侧住宅的日照影响；立面块体图案的划分则依据了原有的城市建筑尺度，从而形成新旧建筑之间的对话。为减少现场施工对密集的周边居民的影响，建筑外墙采用预制混凝土复合挂板，这一措施同时降低了建筑的自重，减少了建筑结构的造价和外墙维护的费用。

摄影：李宁

Hiland · Mingzuo is situated at the trunk roads crossing in Weihai, Shandong Peninsula. Near the coast to the east, it is a real estate building primarily for SOHO units, in addition to commercial functions.

The building is designed with "wind path" that are in the "southwest-northeast" direction to accommodate the dominant southeastern and southern wind in summer and northwestern wind in winter. These paths provide low-tech, simple, and direct natural ventilation from airflow generated by convection and barometric difference. In summer, they direct wind into the building for cooling and dehumidification purpose; and in winter, they minimize the unfavorable effect of wind to the building. Thanks to the CFD (Computational Fluid Dynamics) that was used to simulate, verify, and check the comfort in terms of the wind speed, temperature, and humidity in the design process, maximal number of rooms can be cooled by natural ventilation and less air conditioning devices are necessary in summer. At the inlets of the "wind path" are sealable swing doors. Depending on changes to the exterior weather, the doors are opened or closed to create comfortable microclimate in the building. Additionally, the "wind path" also forms a favorable space for communication and creates excellent platforms for sightseeing to the seashore.

The building intensely dialogs with the urban environment by way of its red pitched roof that not only fits in with the unique architectural style of red-tile pitched roofs in Weihai City but also lets the building minimally affect the sunlight cast on the residential buildings nearby. This new building also dialogs with surrounding buildings by means of its appropriate facade scale. The exterior walls of the building are constructed with composite precast concrete slabs overhung in place so that onsite construction will have minimal effect on the dense surrounding residents and the building weight, the structure cost, and expenses of exterior wall maintenance are reduced.

Photographs: Li Ning

"风径"内部 / Interior of "wind path"

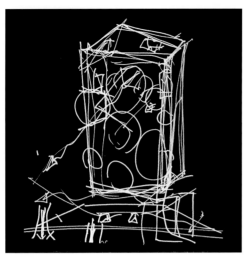

草图 / Sketch

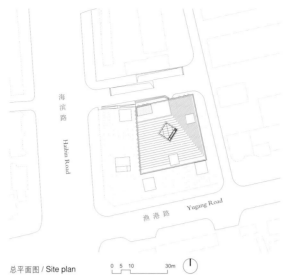

总平面图 / Site plan

"风径"模型研究 / Model study of "wind path"

"风径"生成及自然通风原理分析图 /
Analysis of formation of "wind path" & principle of natural ventilation

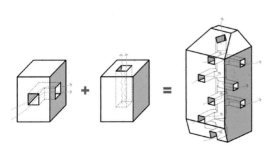
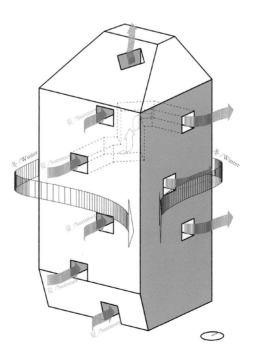

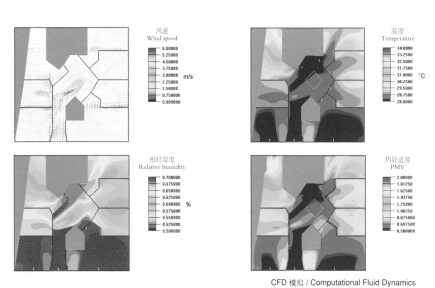

CFD 模拟 / Computational Fluid Dynamics

"风径"模型研究 / Model study of "wind path"

日景 / Day view

夜景 / Night view

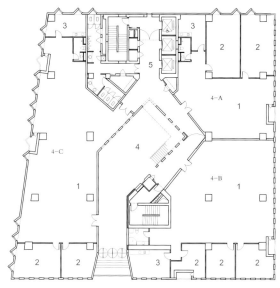

四层平面图 / The 4th floor plan

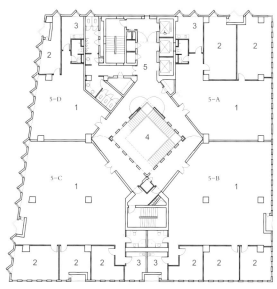

五层平面图 / The 5th floor plan

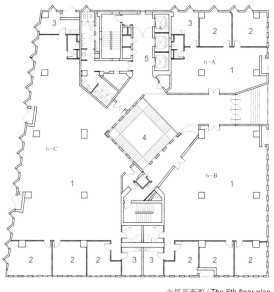

六层平面图 / The 6th floor plan

1	开敞办公区	Open working area
2	办公室	Office
3	服务间	Service
4	"风径"	"Wind path"
5	电梯厅	Lift lobby
6	会所	Club
7	商铺	Shop
8	地下车库	Parking
9	设备用房	Equipments

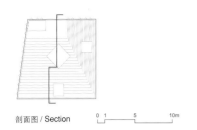

剖面图 / Section

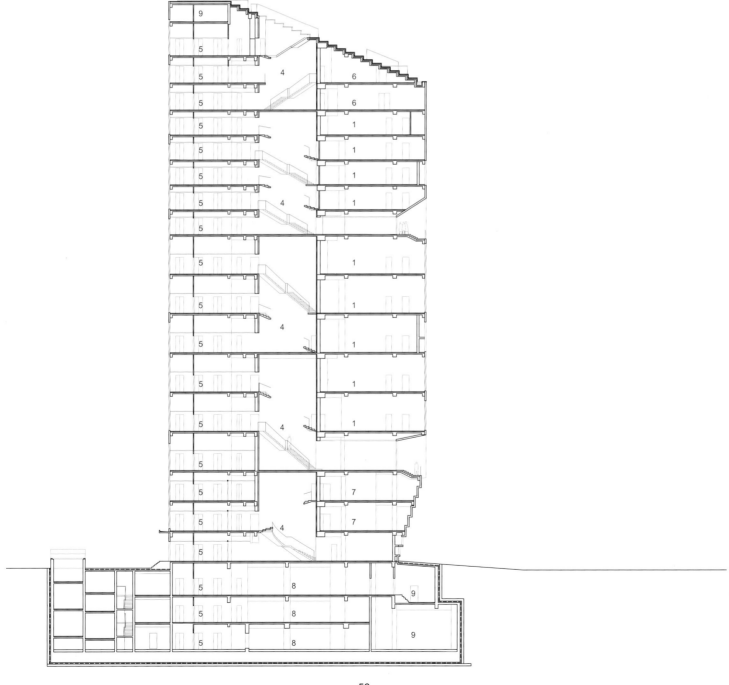

顶层天窗 / Louver of top floor

"风径"内部 / Interior of "wind path"　　　细部 / Detail

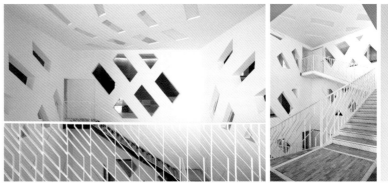

立面局部 / Part of facade

纸砖房　威尼斯
Paper-brick House, Venice
2008

纸砖房位于威尼斯军械库（Arsenale）处女花园，是应第11届威尼斯国际建筑双年展中国馆策展人邀请设计建造的参展作品。以建筑师所在的大型国有设计院日常输出图纸用的纸箱作为"纸砖"砌筑"纸墙"，以日常打印设计图纸剩余的打印机用纸轴（纸管）作为"纸梁"搭建门窗过梁和楼板、屋顶，从而用纸材料建造起一所可供坐卧起居、游戏会客、阅读静思等日常生活的房子。一方面，纸砖房是对大地震中钢筋混凝土建筑及其质量问题使其成为埋葬活人之坟墓的悲剧的直接反应——为什么建筑不能轻一些从而安全一些呢？这也许意味着应对自然的不同方式和建筑另外的发展方向：以柔和的方式应对自然的轻型建筑，而非以抵抗的方式应对自然的重型建筑。另一方面，纸砖房使用了令人目眩的大量图纸箱和打印纸轴，暗示着当下中国生产式输出的建筑设计状态，提示中国建筑师和在中国工作的外国建筑师在应对大量、高速的建设的同时，必须面对的来自质量控制的挑战以及应对策略。

纸砖房的建造方式非常直接：包括"纸箱砖"的强化／防水构造和砌法、"纸管梁"的联结／防水构造和与砌体的联结等。由于现场不能做地下基础的限制，设计了浮置在网眼编织袋（内装级配砂石）上的纸管"筏板"做法，也与威尼斯在海中淤泥层上建造漂浮城市的传统暗合，是一种典型的减震构造。

纸砖房中内向性的庭院空间是建筑的核心，这来自中国的传统；同时建筑也加强了与街道及相邻建筑的关联，并提供外部行人停留休息使用的公共空间和设施，使其具有了更多的公共性。

摄影：张广源、李兴钢、李宁、孙鹏

The architect designed and built this Paper-brick House at the Virgin Garden of Arsenale in Venice for the 11th Venice Biennale of Architecture as invited by the Chinese Pavilion curator. Using paper materials, he built this house for living, entertainment, chatting, reading, and thinking. In fact, the "paper walls" were built with "paper bricks" made from cartons that were daily used to contain building design drawings at China Architecture Design & Research Group, a large state-owned design institute for which the architect works; and the beams, floor slabs, and roof were built with "paper beams" made from wasted paper tubes of printing paper reels used to print design drawings. On one hand, Paper-brick House was a direct response to the tragedy of major earthquakes, in which unqualified reinforced concrete buildings became the tombs of living persons. This may indicate another direction in which buildings might to cope with the nature; that is, lightweight buildings peacefully coexisting with rather than heavyweight buildings. On the other hand, Paper-brick House used large amount of drawing cartons and printing paper reels. This implicated the present status of building production process. It reminded Chinese architects and their foreign peers must overcome the challenges in quality control and must know how to respond to the high-speed construction projects.

Paper-brick House was erected with reinforcement/waterproof construction and installation of "paper carton bricks", connection/waterproof construction of "paper tube beams", and the connection between the "bricks" and the "beams". Because underground foundations could not be built on site, the architect designed a practice whereby paper tube "rafts" were floated on mesh weaving bags that contained graded sand and stone. This practice agreed with the building tradition of Venice that is a floating city built on marshlands.

At the heart of Paper-brick House was an enclosed courtyard space that come from Chinese tradition. With public spaces and facilities for external pedestrians to stay and rest, the building was made more public.

Photographs: Zhang Guangyuan, Li Xinggang, Li Ning, Sun Peng

外景 / View from garden

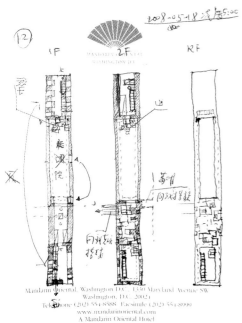

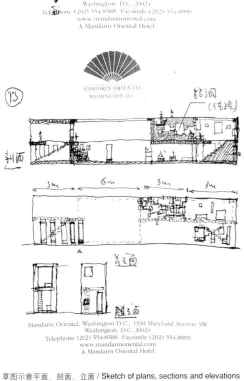

草图示意平面、剖面、立面 / Sketch of plans, sections and elevations

总平面图 / Site plan

1　亭　Pavilion
2　庭院　Yard
3　厅　Living hall
4　入口　Entrance
5　卧室　Bedroom
6　书房　Reading room

轴测图 / Axonometrics　　　轴测平面图 / Axonometrical plans

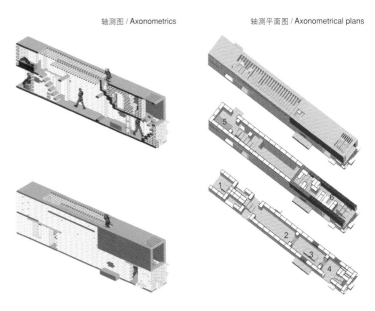

庭院内景 / Interior of yard

由卧室看庭院 / View of yard from bedroom

书房 / Reading room

纸砖与纸管 / Paper bricks & paper tubes

细部 / Details

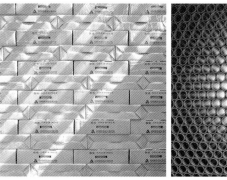 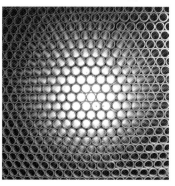

材料 / Materials

材料到达码头 / Materials arriving at jetty

建造过程 / Process of construction

立面局部 / Part of facade

展廊 / Exhibition corridor

李兴钢工作室室内设计　北京

Interior Design of Atelier Li Xinggang, Beijing

2008~2009

李兴钢工作室位于北京中国建筑设计研究院院内一栋小型办公楼的标准层。在有限的造价限制下，怎样使一个原本庸常的结构和空间，在满足使用需求的基础上，变得有趣甚至精彩起来，是设计的重点。

从功能的角度，可以使用到电梯厅、吸烟处、大门、展廊、酒吧、设计工坊、模型制作间、材料样品角、会议／图书室、打印走廊、储藏间、办公室、卫生间、楼梯，等等。从材料的角度，可以观察到压花／普通钢板、欧松板／中密度木板、PVC膜、自流平水泥、镜面、石膏板、软木、硬纸管、图纸箱，等等。从构造做法的角度，可以"阅读"到刚硬的钢板被做成柔和一体的墙地顶曲面交接，柔软如布的膜材成为硬挺的发光顶棚，纸管／纸箱"砌成"的组合墙和坐具，墙面、大门、家具中钢和木的混合／组合使用，大门／旋转镜门的转轴和限位构造，石膏板吊顶的嵌入式照明，以前旧家具的重新利用，门／墙和家具／墙体的一体化设计，等等。从空间游赏的角度，可以体验或想象到似有还无的水面、逆光镂空的亭榭、步移景异的游廊、透露消息的漏窗花墙、看似无用却有用的旋转镜门、狭窄却被放大的吧台、对称却很流动的空间、可以远眺城市风景的楼梯露台，等等；特别是两扇旋转镜门，不仅使光线弯折，更使空间延伸或者旋转，使虚拟与现实空间共存，几乎同等地作用于人的视线、心理和意识。从精神的角度，可以感受到现实与虚幻，实用与无用，物质与意识，理性与感性，秩序与流动……

上述的种种对建筑师自己工作场所的营造、解读和描述，也正大约是建筑师日常工作——思考、言说与实践中的常态：在物质与精神、匠作与哲思、现实与理想的对立统一中煎熬而前行。

摄影：张广源、李兴钢、黄源、张玉婷

Atelier Li Xinggang is situated on a standard floor of an office building. Its design focuses on how to convert an originally plain structure and space into an interesting and even wonderful one at limited cost, while meeting the uses.

The atelier provides functions of lift lobby, smoking pavilion, gate, exhibition corridor, bar, design workshops, model making, material samples corner, meeting room & library, print corridor, storage, office, lavatories, staircase, etc.. The observable materials include embossed & plain steel plates, orientated strand boards (OSBs) & MDF boards, PVC membrane, self-leveling grout, mirrors, plasterboards, cork, hard paper tubes, drawings cartons, etc.. In view of the construction practice, you may "read" curvature joints made from rigid steel plates that softly integrate walls with floors and ceilings; stiff lighting ceilings made from membranes as soft as cloth; composite walls and seats made from paper tubes & cartons; mixed & combined use of steel and wood in walls; embedded lighting in the suspended plasterboard ceiling, etc.. The space also provides an enjoyable sight-seeing scenery where you can experience or imagine a nonexistent pool, a backlit cutout pavilion, a veranda where you can see varied scenery when you step forward, lattice windows and walls through which information can be exposed, staircase and terrace where you can command the city-scope, etc.. Especially, the rotatable mirror-doors not only deflect lights, but also extend or rotate the space to such extent that the virtual and physical space coexist with each other and act on human sight, mind, and consciousness almost to the same degree. Spiritually, you can feel the reality and illusion, utility and inutility, substance and consciousness, sense and sensibility, orderliness and flow…

The architect's construction, interpretation, and description of his own workplace as above may be how he thinks, speaks, and practices in his daily works: struggling and advancing in the unity between substance and spirit, craftsman's working and philosophic thinking, reality and ideality.

Photographs: Zhang Guangyuan, Li Xinggang, Huang Yuan, Zhang Yuting

设计工坊 / Design workshops

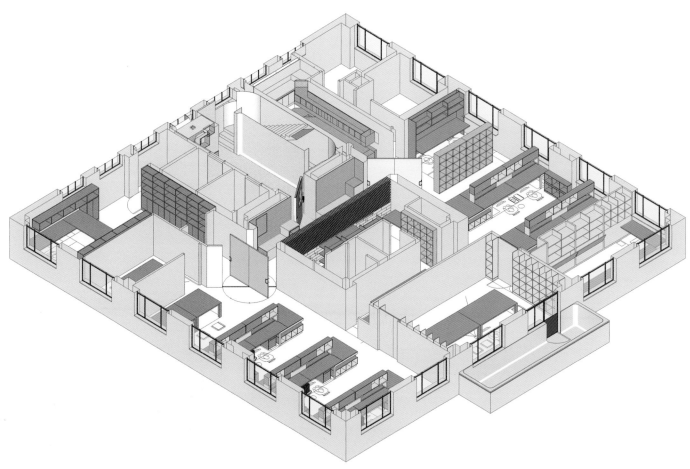

剖轴侧图 / Axonometric

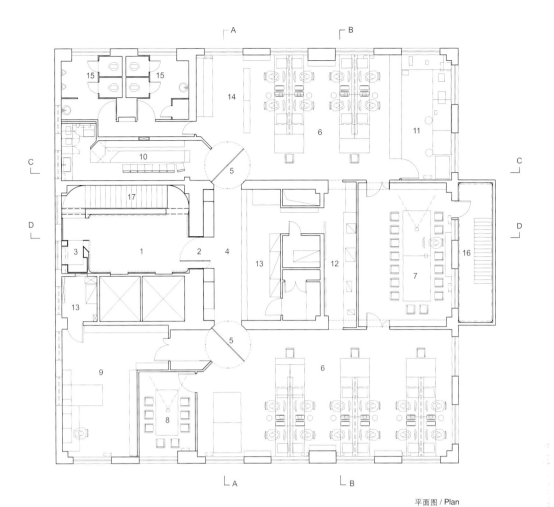

平面图 / Plan

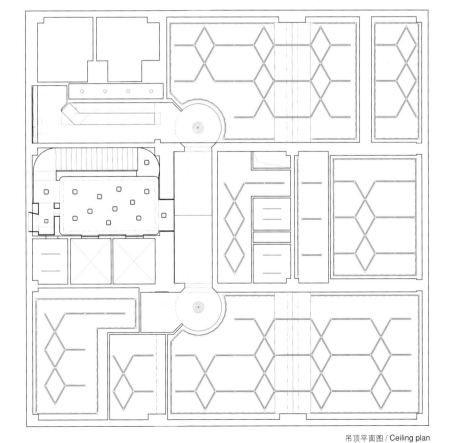

1	电梯厅	Lift lobby
2	大门	Gate
3	吸烟亭	Smoking pavilion
4	展廊	Exhibition corridor
5	旋转镜门	Rotatable mirror-door
6	设计工坊	Design workshops
7	会议图书室	Meeting room & library
8	小会议室	Small meeting room
9	办公室	Office
10	餐吧	Bar
11	模型工作间	Model making
12	打印廊	Print corridor
13	储藏间	Storage
14	材料样品角	Material samples corner
15	卫生间	Lavatory
16	阳台	Bacony
17	楼梯	Staircase

吊顶平面图 / Ceiling plan

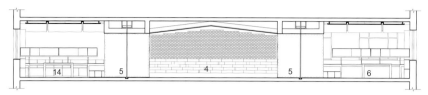

A-A 剖面图 / Section A-A

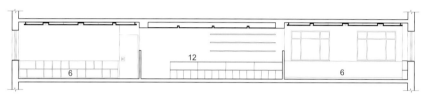

B-B 剖面图 / Section B-B

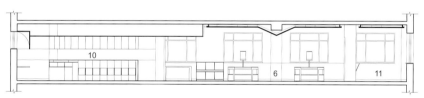

C-C 剖面图 / Section C-C

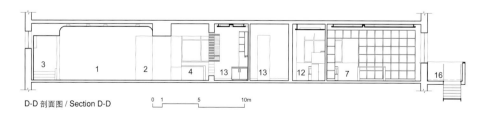

D-D 剖面图 / Section D-D

会议图书室 / Meeting room & library

餐吧 / Bar

餐吧家具轴测图 / Axonometric of furniture of bar

工作桌轴测图 / Axonometric of working-desk

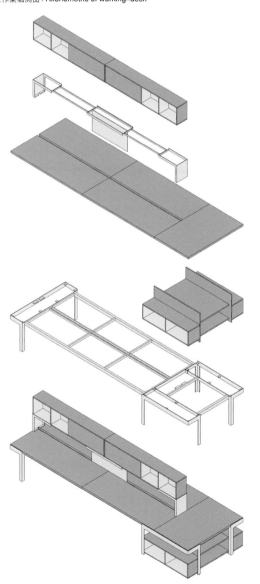

电梯厅 / Lift lobby

办公室 / Office

大门制作 / Making of gate   电梯厅钢板墙焊接 / Welding of steel plate wall of lift lobby   吊顶安装 / Installing of ceiling

大门 / Gate

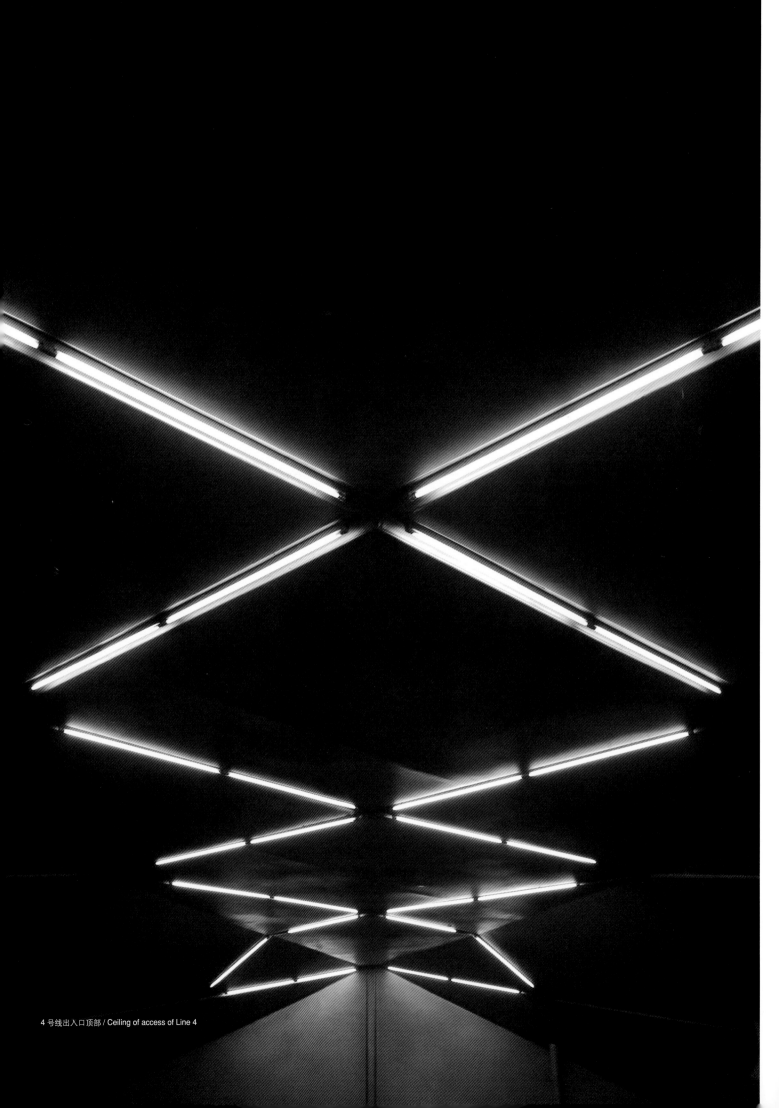

4 号线出入口顶部 / Ceiling of access of Line 4

北京地铁4号线及大兴线地面出入口及附属设施　　北京

Accesses and Subsidiaries of Line 4 & Daxing Line of Beijing Subway, Beijing
2007～2010

地铁4号线是北京市轨道交通网中由南至北穿越了新旧北京城区的轨道交通线，线路全长28.177公里，全程设24座车站。其地面出入口及附属设施的设计需要面对城市空间的特殊性、地面设施类型的复杂性、地下预留站体的结构多变性以及紧张的工期等诸多前提条件。利用钢结构便于标准化、模数化、预制化的特点，将全线出入口站亭设计为在一定模数控制下4种不同规格的系列化网格状钢结构，以适应复杂的地下站体结构尺寸，并在工厂预制标准化的钢结构网格以及与之对应的外墙板块单元，进行现场拼装。在外墙预制板块设计中引入"城市画框"的概念，运用金属板、彩釉印刷玻璃和透明玻璃的不同组合，对变化的城市环境进行多样性摄取，使用者可以透过取景窗辨识出入口所处城市空间的典型特征。建筑形体采用了坡形山墙断面和矩形断面过渡的基本形式，凸显地铁出入口的功能性与标识性，同时呼应了4号线串联旧城和新城的文脉关系，使地铁出入口站亭成为市民了解城市文化，体验城市空间，小建筑中见大意境的公共交通建筑。

地铁大兴线作为4号线的南延线，正线全长21.756km，新设11座车站，全线贯穿大兴区生活商业中心与南部新兴产业园区，处于由现代城市景观向郊区自然景观过渡的特定城市空间之中。其地面出入口及附属设施依然延续了4号线坡形断面与矩形断面过渡的基本形式。为了突出本线的地理文化特征，呼应4号线的"城市画框"，引入了"城市画卷"的设计概念，运用现代彩釉玻璃丝网印刷技术，将元代画家倪瓒的《山水图》和宋代画家王希孟的《千里江山图》经过抽象后拓印于玻璃幕墙之上，唤起对大兴地区自古而成的自然山水格局的回忆。整幅画面采用5种不同透明度的白色釉块组合抽象表现，虚实相叠，并利用白天自然光线和夜晚室内光线的变化，形成对玻璃幕墙画面多角度、多时段的不同解读，同时也形成与中国传统山水画相对应的远观其势，近观其质的视觉传承。乘客穿越出入口站亭空间出入城市，视线会被玻璃帷幕的光影变化吸引，从而获得历史与现代时空交织的独特体验。

Subway Line 4 runs through a 28.177 km-long route connecting 24 stations from the south to the north across the new and old urban areas of the city. The design of its accesses and ancillary facilities has to meet many prerequisites, such as the particularities of urban spaces, complex types of ground facilities, varied structures of the reserved underground stations. So, all the accesses adopt a series of steel lattice structures, which can be easily standardized, modularized, and prefabricated. Standardized lattices and their corresponding exterior wall boards are prefabricated in manufactories and assembled on site. An "urban picture frame" concept is introduced. Different combinations of metal plates, color glazed glass, and transparent glass capture different aspects of the changing urban environments. The accesses are generally profiled as the transition between pitched-gable and rectangular sections to identification as subway accesses and to suit the Line 4 that is a cultural vein connecting the old to the new urban areas. Thus, these accesses have become public transport buildings of significance, i.e., for citizens to learn about the culture and experience the space of the city.

As the southern extension of Line 4, the Subway Daxing Line passes by 11 new stations along its 21.756 km-long main route across the living and commercial center of Daxing District. It is located in a special urban space transiting from modern city landscape to the natural suburban landscape. The design adopt the general transition between pitched-gable and rectangular sections of Line 4. They have been designed with an "urban painting" concept to highlight the geographical and cultural characteristics. The "Mountain and Water" painted by Ni Zan, a Chinese painter in the Yuan Dynasty, and the "Thousands Miles of Rivers and Mountains" painted by Wang Ximeng, a Chinese painter in the Song Dynasty are abstracted and inscribed on glass curtain walls. The entire picture is abstractly presented by combining of five white glaze blocks of different transparency. Thanks to the overlapping of virtual and real elements and the variation from natural light in daytime to the interior light at night, this picture on the glass curtain wall can be interpreted in different manners depending on the viewpoints and time periods. Passengers will be attracted by the changing light and shadow of the glass curtain and will experience the interlacing of the past and the present, the time and the space.

摄影：张广源、邱涧冰、李喆

Photographs: Zhang Guangyuan, Qiu Jianbing, Li Zhe

4号线国家图书馆站西南口 / Southwest access of National Library station of Line 4

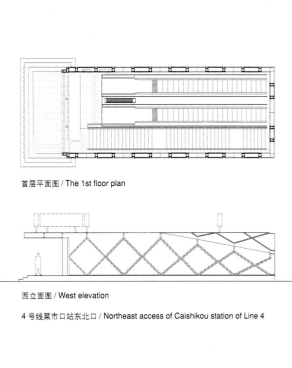
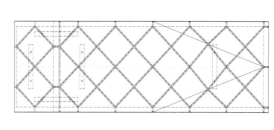

首层平面图 / The 1st floor plan　　　　　　　　　　　　　　　　　屋顶平面图 / Roof plan

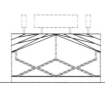
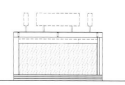

西立面图 / West elevation　　　　　　　　　　　　　　　　　　　南立面图 / South elevation　　　北立面图 / North elevation

4号线菜市口站东北口 / Northeast access of Caishikou station of Line 4

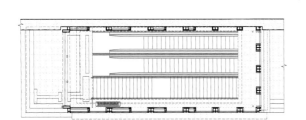
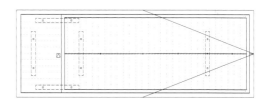

首层平面图 / The 1st floor plan　　　　　　　　　　　　　　　　　屋顶平面图 / Roof plan

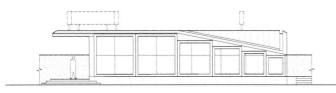

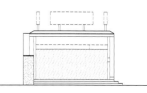

南立面图 / South elevation　　　　　　　　　　　　　　　　　　　西立面图 / West elevation　　　东立面图 / East elevation

4号线动物园站北口 / North access of Beijing Zoo station of Line 4

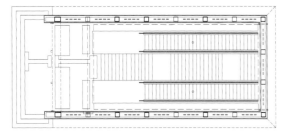
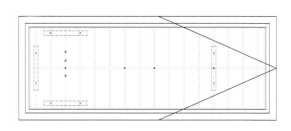

首层平面图 / The 1st floor plan　　　　　　　　　　　　　　　　　屋顶平面图 / Roof plan

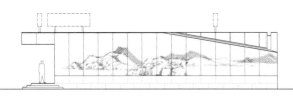
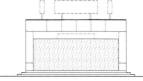

东立面图 / East elevation　　　　　　　　　　　　　　　　　　　南立面图 / South elevation　　　北立面图 / North elevation

大兴线枣庄站东北口 / Northeast access of Zaozhuang station of Daxing Line

4号线西四站西北口 / Northwest access of Xisi station of Line 4

屋面 Roof:
盒型金属板面层 Box type sheet metal finish
柔性防水透气膜防水层 Flexible breathable waterproof membrane
1.2厚铝板找坡层 Aluminum plate to falls t=1.2
50厚岩棉隔热层 Rock wool t=4
1.2厚铝板垫层 Aluminum plate t=1.2
50C型轻钢龙骨 50C Metal stud

加强方钢，沿折线方向板块内通长
Reinforced square steel, full length along the line direction in the plate

4mm厚镀锌钢槽 Galvanized steel tank t=4
50mm厚岩棉 Rock wool t=4

屋面 Roof:
盒型金属板面层 Box type sheet metal finish
柔性防水透气膜防水层 Flexible breathable waterproof membrane
1.2厚铝板保护层 Aluminum plate t=1.2
50厚岩棉隔热层 Rock wool t=4
1.2厚铝板垫层 Aluminum plate t=1.2
50C型轻钢龙骨 50C Metal stud

1.2mm厚铝板 Aluminum plate t=1.2

灯具 Lamp

2厚铝单板 Aluminum veneer t=2
3厚钢板 Steel plate t=3

顶棚 Ceiling:
50C型轻钢龙骨 50C metal stud
1.2厚铝板垫层 Aluminum plate t=1.2
盒型金属板面层 Box type sheet metal finish

钢板 3厚 Steel plate 3.0mm
铝单板 2厚 Aluminum veneer 2.0mm

加强方钢，沿折线方向板块内通长
Reinforced square steel, full length along the line direction in the plate

外墙面 External wall:
盒型金属板面层 Box type sheet metal finish
柔性防水透气膜防水层 Flexible breathable waterproof membrane
1.2厚铝板保护层 Aluminum plate t=1.2
50厚岩棉隔热层 Rock wool t=4
1.2厚铝板垫层 Aluminum plate t=1.2
50C型轻钢龙骨 50C Metal stud

内墙面 Wall:
50C型轻钢龙骨 50C Metal stud
1.2厚铝板垫层 Aluminum plate t=1.2
盒型金属板面层 Box type sheet metal finish

密封胶 Sealant

夹胶玻璃，内表面作彩釉印刷
Laminated glass, colour glaze printing inside

25厚防火涂料 Fireresistant coating t=25

外墙面 External wall:
盒型金属板面层 Box type sheet metal finish
钢衬板 Steel liner
柔性防水透气膜防水层 Flexible breathable waterproof membrane
1.2厚铝板保护层 Aluminum plate t=1.2
50厚岩棉隔热层 Rock wool t=4
1.2厚铝板垫层 Aluminum plate t=1.2
50C型轻钢龙骨 50C Metal stud
加强肋 Reinforced rib

内墙面 Wall:
50C型轻钢龙骨 50C Metal stud
1.2厚铝板垫层 Aluminum plate t=1.2
盒型金属板面层 Box type sheet metal finish

补做埋件 Embedment

结构支架 Structural support

原装修面层 Original surface decoration

石材墙面 Stone wall

原结构 Original structure

原防水层 Original waterproof layer

墙身及屋顶详图 / Details of wall & roof

4号线角门西站出入口施工 /
Construction of access of Jiaomenxi station of Line 4

4号线灵境胡同站出入口施工 /
Construction of access of Lingjing Hutong station of Line 4

4号线国家图书馆站西南口 / Southwest access of National Library station of Line 4

4号线北宫门站东南口 / Southeast access of Beigongmen station of Line 4

4号线中关村站出入口施工 / Construction of access of Zhongguancun station of Line 4

4号线魏公村站出入口施工 / Construction of access of Weigongcun station of Line 4

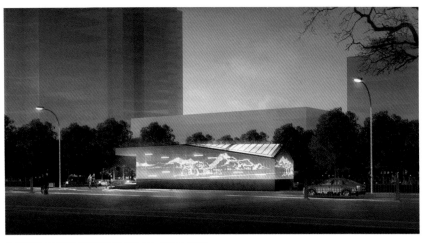

大兴线出入口效果图 / Perspective of access of Daxing Line

88

大兴线天宫院站西北口与东北口 / Northwest & northeast accesses of Tiangongyuan station of Daxing Line

立面局部 / Part of facade

北京地铁昌平线西二旗站 北京

Xi'erqi Station of Changping Line of Beijing Subway, Beijing
2008 ~ 2010

　　西二旗站是北京地铁昌平线的南起点站，是与城铁13号线的换乘站。车站形式为半地下一层、地上二层，四柱三跨框架式高架车站。其中的昌平线部分为高架侧式站台（二层），城铁13号线部分为地面侧式站台（一层），站厅位于地面一层。根据侧式站台的特征，车站采用了双四边形组合建筑断面形式，并以PTFE膜结构作为屋面和立面维护材料和结构，实现了内部基本无柱的长向大空间。

　　适应地铁交通需求的双四边形组合筒状空间被直接外现为建筑的造型，并采用了折纸状的结构和形式，实现了模数化、标准化、预制化的设计与建造，简洁、流畅而富于韵律感，真实而充分地体现了张拉膜材料和结构的特征，建筑形式也与结构、排水等要求相适应。车站的出入口大厅、雨篷等元素采用了一体化的建筑语汇。

　　由于PTFE膜材料的半透光特性，白天的车站内可以透射进柔和而充分的自然光线，无须人工照明；到了夜晚，车站内的灯光透射出去，使得车站就像两条发光的纸灯笼，乘坐地铁回家的人远远就可以看到。地铁西二旗站以其不动声色的建筑姿态与人们日常生活的密切关联并成为重要的"城市标志物"。

Xi'erqi Station is the southern terminal of Subway Changping Line in Beijing and its transfer station to Subway Line 13. The station is an elevated frame-type building consisting of a semi-underground storey and two above-ground stories supported by four colonnades that create three spans. The Changping Line part consists of an elevated side platform on the second storey and the Line 13 is part of a ground side platform on the ground storey. Based on the characteristics of side platforms, the cross-section of the buildings consists of two quadrangles combined with each other. PTFE membrane functions as the material and structure that enclose the roof and facade to provide a long-large space in which few columns are present.

Accommodating subway transportation, the tubular space formed by the combination of the two quadrangles is directly exposed to show the shape of the architecture. The structure and form, like something folded with paper, enable modular, standardized, prefabricated construction. The concise, fluent, and rhythmic architecture indeed expresses the characteristics of tensile membrane. The architectural form also meets the structure and drainage requirements.

Because of the translucence of the PTFE membrane, soft and sufficient natural light can be transmitted into the station and no artificial lighting is necessary in daytime. At night, the lamp light transmitting out of the building makes the station like two bands of lighting paper lanterns that can be seen from far away by subway commuters going home. With its quiet architectural pose but close relationship with our daily life, the station becomes an important "city landmark".

摄影：张广源、张音玄、黄达达

Photographs: Zhang Guangyuan, Zhang Yinxuan, Huang Dada

夜景 / Night view

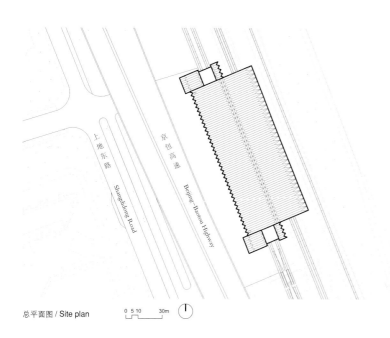

总平面图 / Site plan

西侧外景 / West exterior

13号线站台内景 / Interior of platform of Line 13

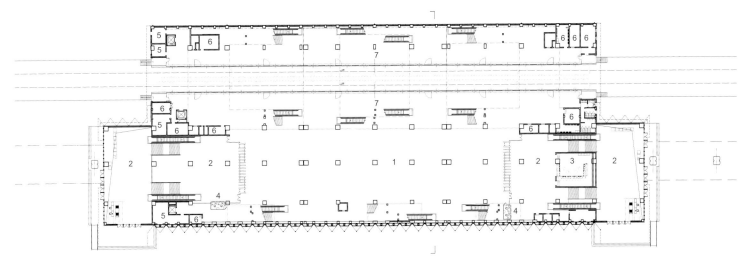

昌平线站厅层及13号线站台层平面图 / Plan of entrance hall of Changping Line & platform of Line 13

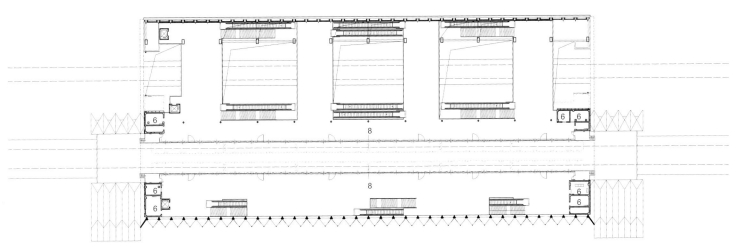

昌平线站台层平面图 / Plan of platform of Changping Line

1	付费区	Paid area
2	非付费区	Unpaid area
3	合用车站控制室	Shared station control room
4	客服中心	Passenger centre
5	办公	Office
6	设备用房	Equipments
7	13号线站台	Platform of Line 13
8	昌平线站台	Platform of Changping Line

室内天桥 / Inner bridge

楼扶梯 / Stair & escalator

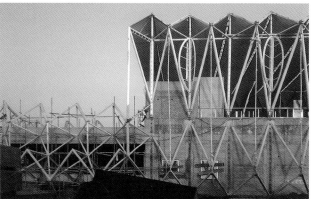
膜结构施工 / Construction of tensile structure

客流模拟 / Simulation of passenger flows

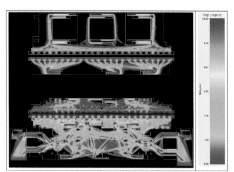
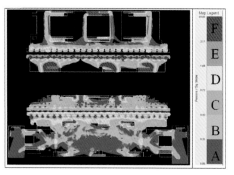

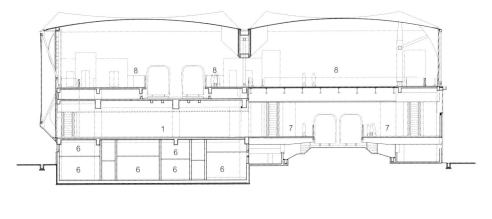

剖面图 / Section

室内天桥 / Inner bridge

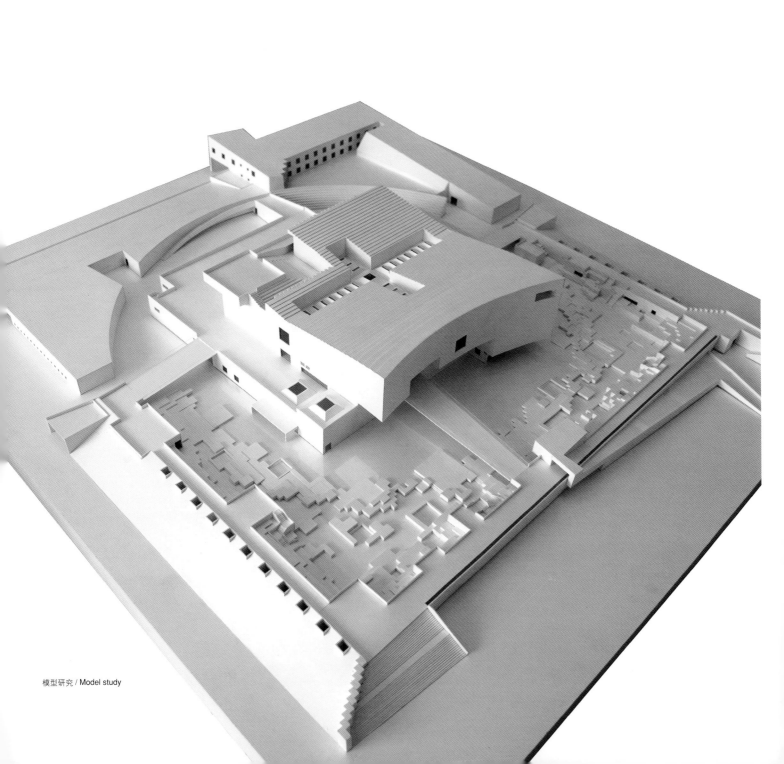

模型研究 / Model study

商丘博物馆   商丘

Shangqiu Museum, Shangqiu

2008 至今

商丘博物馆位于商丘西南城市新区，收藏、陈列和展示商丘的历代文物、城市沿革和中国商文化历史。博物馆主体由三层上下叠加的展厅组成，周围环以景观水面和庭院，水面和庭院之外是层层叠落的景观台地和更外围高起的堤台（下面设室外展廊），文物、业务和办公用房组成L形体量，设置于西北角堤台之上，设南北东西四门。博物馆的整体布局和空间序列是对商丘归德古城为代表的黄泛区古城池典型形制和特征的呼应与再现，博物馆犹如一座微缩的古城。上下叠层的建筑主体喻示"城压城"的古城考古埋层结构，也体现自下而上、由古至今的陈列布局。

参观者由面向阏伯路的大台阶和坡道登临堤台，沿引桥凌水由中部序言厅入"城"，自下而上沿坡道陆续参观各个展厅，最后到达屋顶平台，由建筑各角不同方向的眺望台与著名古迹——阏伯台、归德古城、隋唐大运河码头遗址等遥遥相望，怀古思今。

周围下沉式的景观台地对考古现场的模拟，使得建筑主体犹如被发掘出来；古象形文"商"字的涵义是"高台上的子姓族人"，博物馆所形成的层层高起的堤岸、平台和其上的参观者将真正再现"商"字的古老渊源和意象，把历史和现在绝妙地关联为一体。

摄影：李兴钢、李喆

Shangqiu Museum is situated in the southwestern urban area of Shangqiu City, Henan Province. It collects, displays the cultural relics of Shangqiu City, the evolution of the city, and the culture and history of "Shang" Dynasty. The main body of the museum consists of three overlapped stories of exhibition halls surrounded by water landscape and a courtyard, beyond which are tiers of landscape platforms under which is an outdoor exhibition corridor. Like a miniature ancient city, the museum has the overall layout and spatial sequence that correspond to the ancient Gui'de city of Shangqiu, which represents the typical format and characteristics of ancient cities in the region formerly flooded by the Yellow River. The overlapped main body of the building reflects the archaeological buried structure in which one town is on the top of another.

Visitors may climb onto the levee along the giant steps and slope facing E'bo road. Then they can pass the water over an approach bridge and enter the "city" through the preface hall. After visiting the halls along the slope from the bottom up, they will reach the roof platform, where there are gazebos at the corners to command the famous historic spots, such as the E'bo Platform, the ancient Gui'de city, and the ruins of the Grand Canal of the Sui and Tang Dynasties.

The surrounding sunken landscape platforms simulate the excavation site of cultural relics. The ancient pictograph "商" (Shang) means "members of the clan with a family name of Zi standing on a high platform", which is symbolized by visitors standing on the tiers of levee and platforms. This indeed reproduces the origin of the character and elegantly combines the past and the present.

Photographs: Li Xinggang, Li Zhe

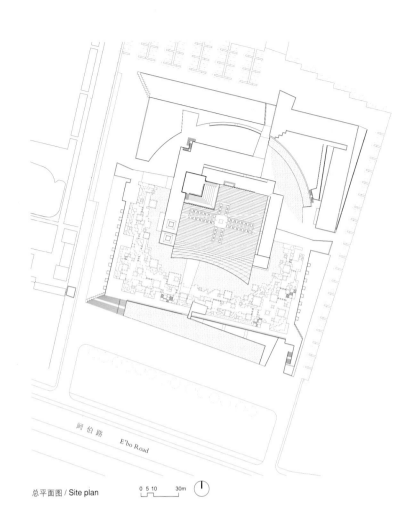

总平面图 / Site plan

模型研究 / Model study

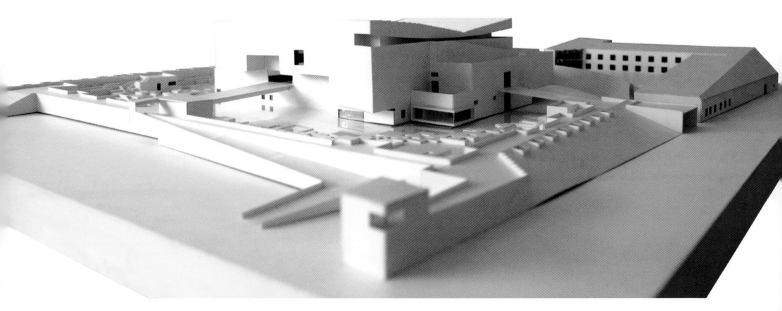

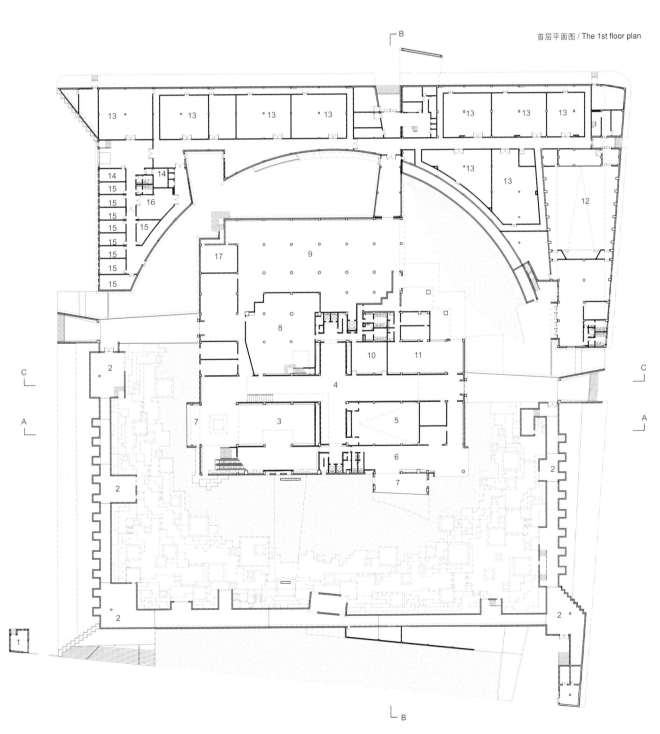

1	警卫室	Guard
2	室外展廊	Outdoor exhibition corridor
3	序言厅	Preface hall
4	共享大厅	Main hall
5	放映厅	Cinema
6	茶餐厅	Tea restaurant
7	亲水平台	Waterside terrace
8	展厅	Exhibition hall
9	多功能大厅	Multi-functional hall
10	信息厅	Information room
11	贵宾区	VIP area
12	报告厅	Lecture hall
13	展品库房	Storage
14	安保室	Security control
15	研究用房	Research room
16	消毒室	Disinfection room
17	设备用房	Equipments

首层平面图 / The 1st floor plan

施工现场 / Construction site

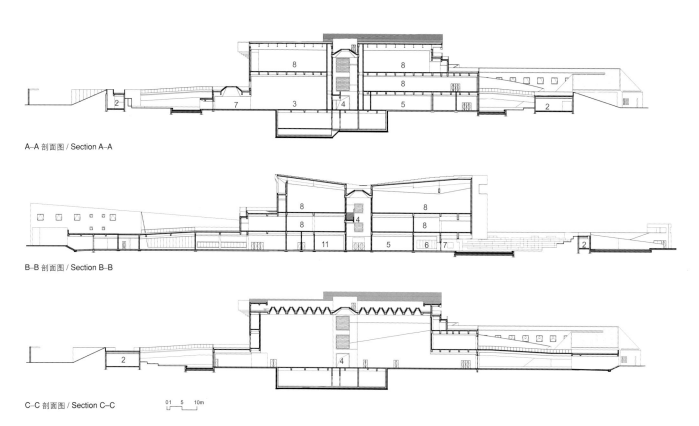

A–A 剖面图 / Section A–A

B–B 剖面图 / Section B–B

C–C 剖面图 / Section C–C

施工现场 / Construction site

天窗 / Louver

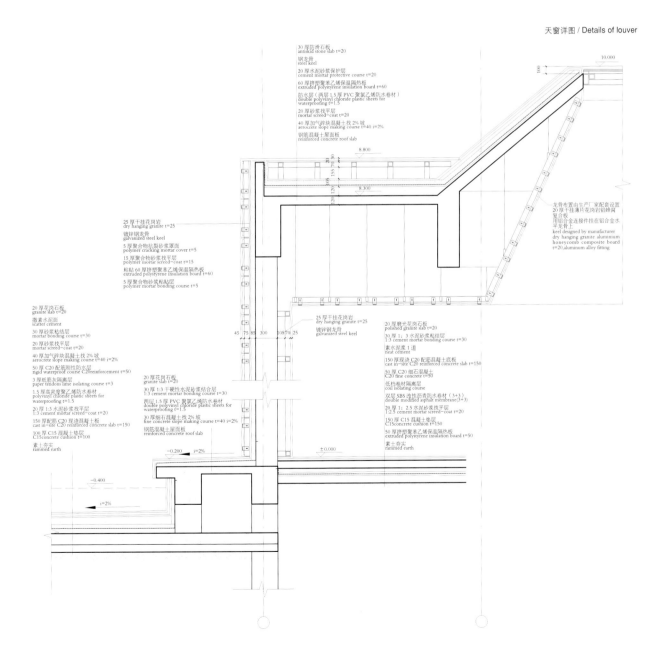
天窗详图 / Details of louver

模型研究 / Model study

"第三空间" 唐山
"The Third Space", Tangshan
2009 至今

"第三空间"综合体位于唐山市建设北路东侧。用地西临城市干道、东临平行排布的多层住宅楼群，南北狭长。包括南北两栋百米塔楼（会所单元）和下部连为一体的公共裙房（门厅、商铺、超市、餐厅、泳池、SPA、酒吧、歌舞厅等公共服务休闲设施）。

设计构思源于勒·柯布西耶的"别墅大厦"，建筑的主体功能及空间设计均体现"第三空间"的定位和理念，即在城市中心地带提供76套（500～1000平方米／套）介于办公与居住性质之间的"私人会所"，满足城市企业领袖这一特定人群的特定需求：身处城市的繁华之中却能静心养神、决策思考、待客会谈。每套"会所"内部错层设计，不同功能处于不同标高，并利用南北立面的景观亭台或顶层的室外庭院，形成静谧的空间氛围和丰富的空间转换。建筑的形体及其布局完全源于针对东侧住宅严苛的日照分析结果（新建建筑不能对住宅日照产生不利作用和影响）——即塔楼与南北向主干道的成35°（而非90°）角布置，和裙房的三段斜顶体量。各套复式私人会所在朝向东南的立面悬挑出不同尺度及方向的室外亭台，形成繁复密匝的"空中建筑群"的意向，并使建筑的空间和形象与城市景观产生因借和互动。上部的两个塔楼和下部的裙房之间各设一个通透的空中大堂，既有开阔的视野，又凸现了建筑上部形体的悬浮感。

主体立面材料为白色仿清水混凝土GRC挂板，立面材料延伸至屋面，形成完整的、具有雕塑感的建筑形体。东南向悬挑的空中亭台内表面采用当地特产的不规则"唐陶"彩色瓷片拼砌，在沉稳厚重的建筑主体气质之下，形成较为独特的个性表达。

摄影：孙鹏

"The Third Space" is a complex situated to the east of the Jianshebei Road in Tangshan City. The complex includes two 100-meter-high towers used as units of "private club", one in the north and the other in the south. At the lower part of the complex is a public annex joining the towers together, providing utilities and entertainment facilities like entrance hall, shops, supermarket, restaurant, swimming pool, SPA, bar, and ballroom.

With the design concept originated from Le Corbusier's "Villa Mansion", the building has its main function and spatial design in line with its concept of "The Third Space" - to provide 76 units of "private club" somewhat between office and residential function with the area of 500~1,000m^2 per unit in the downtown area, enabling the entrepreneurs of the city to relax, think, make decisions, and consult with visitors in a quite space within crowded city. Each unit of club adopts a split-level interior design whereby different functions are deployed to different elevations, and is equipped with sightseeing pavilions on the southern and northern facades. The shapes and layout of the buildings have been determined based on a strict sunlight analysis to prevent it from negatively affecting the residence nearby. The towers are arranged in a 35-degree-angle (rather than 90-degree) with the north-south trunk road and the annex is a three-section mass covered by pitched roofs. On the southeastern facade of each duplex club are exteriorly suspended pavilions of different scales and directions. These pavilions form an imagery of a dense "group of aerial buildings". They also enable the space to suit and borrow scenery from the urban landscape. A transparent sky lobby between the lower annex and each of the two upper towers provides an open view and creates a floating feel of the upper body of the building.

The facade of the main body is made of white as-cast-finish-concrete GRC hanging plates, which extend to the roof and form a complete architectural volume. Colorful ceramic chips of irregular shapes locally made in Tangshan are attached to the internal face of the aerial pavilions facing southeast.

Photographs: Sun Peng

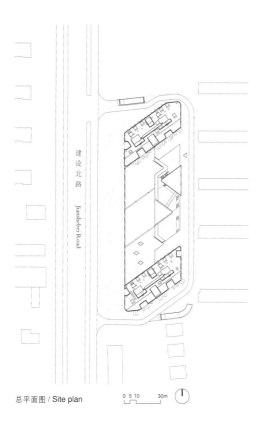

总平面图 / Site plan

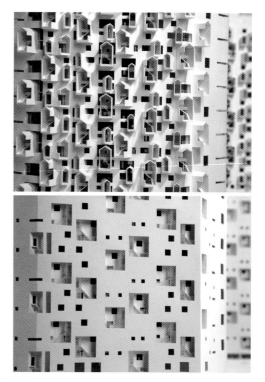

模型研究 / Model study

施工现场 / Construction site

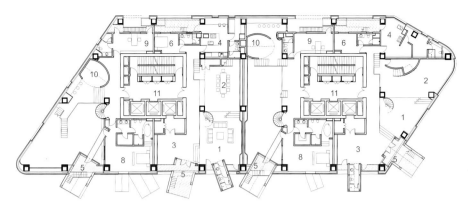

北楼七层平面图 / The 7th floor plan of north building

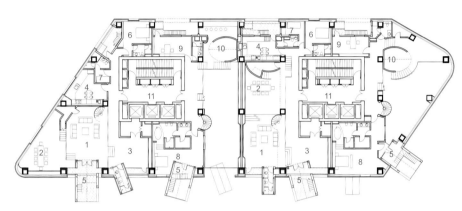

北楼八层平面图 / The 8th floor plan of north building

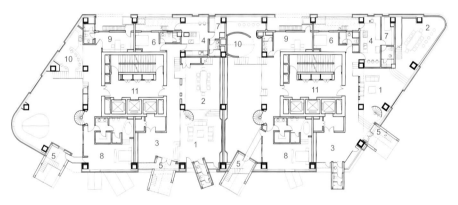

南楼七层平面图 / The 7th floor plan of south building

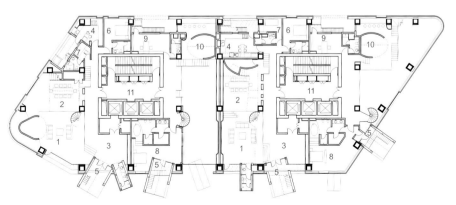

南楼八层平面图 / The 8th floor plan of south building

1	客厅	Living room
2	餐厅	Dining
3	门厅	Entrance
4	厨房	Kitchen
5	景观亭台	Sightseeing pavilion
6	客卧	Guest room
7	佣人房	Servant room
8	主卧	Master room
9	书房	Reading room
10	红酒收藏	Wine collection & bar
11	电梯厅	Lift lobby
12	影视厅	Video room
13	游泳池	Swimming pool
14	空中大堂	Sky lobby
15	餐厅	Restaurant
16	Spa 会所	Spa
17	酒吧	Bar
18	KTV	KTV
19	高级超市	Supermarket
20	茶室	Tea
21	设备用房	Equipments
22	地下车库	Parking

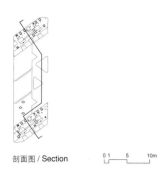

剖面图 / Section

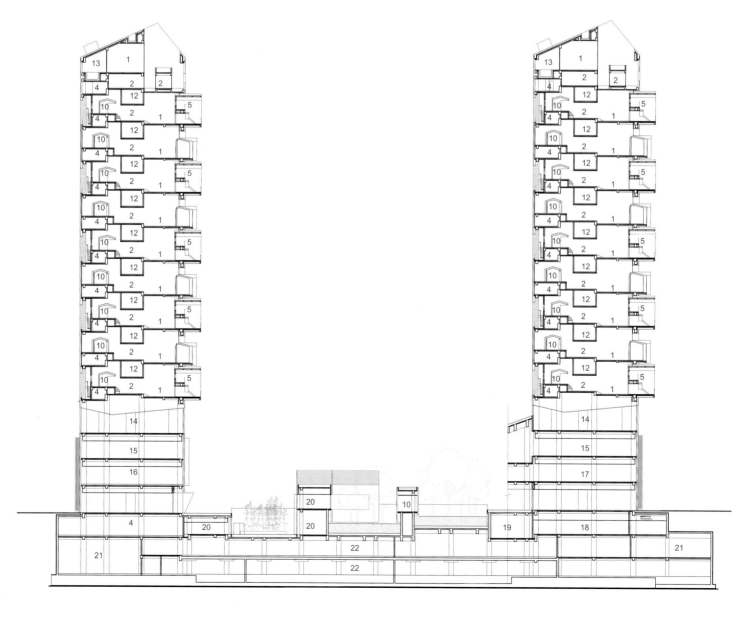

施工现场 Construction site

模型研究 / Model study

元上都遗址博物馆　正蓝旗
Museum for Site of XANADU, Zhenglanqi

2009 至今

元上都遗址是中国元代北方骑马民族创建的一座草原都城的遗迹，位于内蒙古自治区锡林郭勒盟正蓝旗五一牧场境内、闪电河（滦河上游）北岸冲积平地上，山川雄固，草原漫漫，层叠深远，尤其适合登高远眺。元上都遗址博物馆是配合元上都古都城遗址申报世界文化遗产的配套项目。主要功能包括展厅、观众服务、藏品库房、内部办公、考古科研等。

建筑选址于元上都遗址南向5公里，乌兰台的东侧面向遗址方向的半山腰处，参观者由南而来，绕山而行，通过北侧山脚下的道路进入博物馆区，有豁然出现之感。设计结合并充分利用现状废弃的采矿场来布置博物馆的建筑主体，以修整被采矿破坏的山体。博物馆工作人员入口设置在现状的一处折线形采矿条坑南端，并将办公用房沿折线凹地布置，且沿山坡形状覆土；保留另一处现状圆形矿坑，修整作为博物馆的下沉庭院，观众服务区环绕着此庭院。遵循对文化遗产环境完整性的最小干预原则，将巨大的建筑体量掩藏在山体之内，仅半露一小段长条形体，并将其指向都城遗址中轴线上的起点——明德门，使建筑对遗址有理想的视角和轴线关联；而由明德门处看遗址博物馆，建筑则缩为一个隐约的方点，体现出对遗址环境完整性的尊重以及人工与自然的恰切对话和协调。

沿着博物馆的内外参观路径设置了一系列远眺遗址和草原丘陵地景的平台，直至到达山顶敖包，长长的路径和不断停驻的平台是博物馆不可分割的组成部分，将元上都的历史、文化和景观在此串联。

Site of XANADU is the ruin of the Yuan Dynasty Upper Capital built by the horse-riding Mongolians in northern China. This capital was situated in the Wuyi Pasture, Zhenglanqi of Xilin Gol League in the Inner Mongolia Autonomous Region. Museum for Site of XANADU is an auxiliary project that helps make the ruin a World Heritage Site. It provides functions like exhibition halls, visitors' service, storage of collections, offices, and archeological research, etc.

The museum is located 5 km south of the Site of XANADU ruins, and on the half way up the ruin-facing eastern slope of Wulantai Mountain. Visitors coming from south will walk around the mountain before they catch a sudden sight of the museum by accessing the museum area along a road at the northern foot of the mountain. The staff entrance of the museum is deployed at the southern end of a polyline-shaped mining pit. Offices are arranged in the valley of the polyline and covered with earth to fit in with the shape of the slope. Another circular mining pit is retained and modified into a sunken courtyard of the museum surrounded by the visitors' service facilities. To minimize intervention to the integrity of the cultural heritage, all of the huge architectural mass except for a short-strip-shaped section is hidden in the mountain. Moreover, this section is half-exposed and points at Mingde Gate, the starting point of the axis of the Site of XANADU. In such a way, the architecture is ideally related to the ruin in terms of the view angle and axis. If one views from Mingde Gate, the museum becomes an indistinct square dot. All these reflect the respect to the integrity of the ruin and the appropriate dialog between manmade and natural elements.

Platforms provide commanding views over the ruin, the grassland, and hills all the way to the top of the Wulantai Mountain. The long paths and lingering platforms are integral parts of the museum that connect the history, culture and landscape of XANADU.

摄影：张广源、李兴钢、邱涧冰

Photographs: Zhang Guangyuan, Li Xinggang, Qiu Jianbing

由遗址远眺乌兰台 / Look far to Wulantai Mountain from the ruin

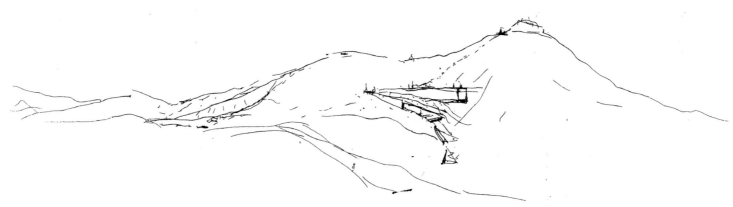

草图 / Sketch

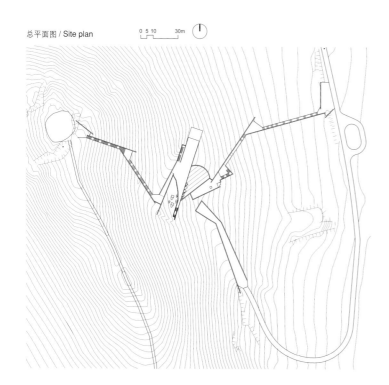

总平面图 / Site plan

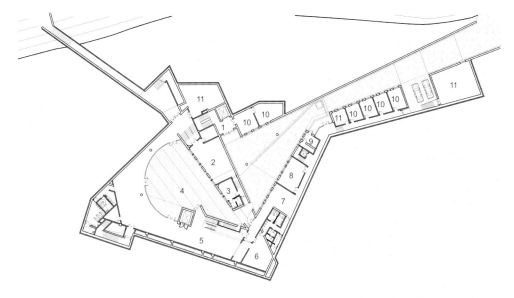

地下一层平面图 / The basement floor plan

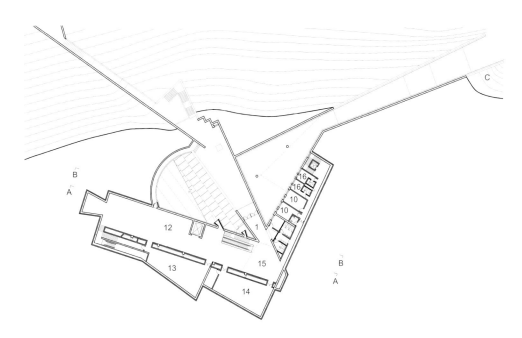

首层平面图 / The 1st floor plan

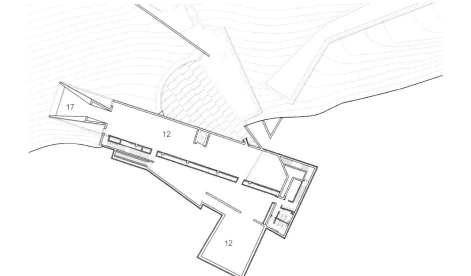

二层平面图 / The 2nd floor plan

1	门厅	Entrance
2	多功能厅	Multi-function hall
3	休息室	Break room
4	室外庭院	Courtyard
5	观众服务	Spectator service
6	厨房	Kitchen
7	食堂	Dining hall
8	会议室	Meeting room
9	消防控制室	Fireproof control room
10	办公	Office
11	设备用房	Equipments
12	主题展厅	Theme exhibition hall
13	临时展厅	Temporary exhibition hall
14	4D 影院	4D cinema
15	序言厅	Preface hall
16	宿舍	Dormitory
17	观景厅	View hall

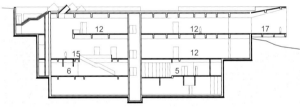

A-A 剖面图 / Section A-A

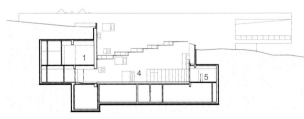

B-B 剖面图 / Section B-B

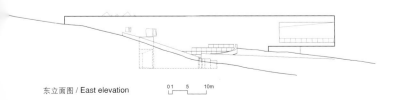

东立面图 / East elevation

施工现场 / Construction site

由博物馆远眺遗址 / Look far to the ruin from the site   赭红的山岩 / Siena red mountain rock

施工中的屋顶 / Roof in construction

海南国际会展中心　　海口
Hainan International Conference & Exhibition Center, Haikou
2009 ~ 2011

海南国际会展中心位于海口市西部，新的城市组团北部，用地北侧滨海，周围相邻酒店、公寓、城市公园等。包括展览中心和会议中心两大功能及其相关附属设施。

建筑主体采用一体化的设计手法，将展览中心和会议中心整合处理为一个巨大的完形体量，匍匐在南中国海滨，成为城市南北向景观轴线的结束与高潮。作为新城中心区轴线的底景，这一庞大的建筑有着明确的、与新城中心合一的轴线，同时又因地形、海岸和内部功能的不同而产生了丰富的非对称建筑形体，成为在规则韵律中富含变化、整体造型协调统一的一个巨型建筑体。建筑造型是处于"像与不像"之间的抽象人工物，"海水"、"云团"、"沙滩"、"海洋生物"、"海上景象"……似是而非，但总与大自然海洋气质相契合。规律起伏波动的屋面壳体钢结构由等截面的密格式布置的圆形钢管梁（中部区域）和钢网架（边缘区域）组合而成，以工厂预制、现场组装、吊装就位的方式建造，既塑造出独特的建筑外观，也与室内空间高度统一，营造出适应不同规模和使用需求的展览和会议、休闲、服务等空间。壳顶单元的天窗和高侧窗为室内空间提供了充足的、不断变幻角度的自然光线。环绕整个建筑外侧的半拱形檐廊、雨棚及其下结合窗台设计的"石凳"，是应对当地强烈的日晒条件，为室外的使用者提供的遮阳和休息设施。

采用的建筑材料有石材、曲面铝板、现制GRC及屋面涂料等。大部分室内钢结构完全露明，呈现出结构自身的美感和韵律。

摄影：张广源、李兴钢、张玉婷

Hainan International Conference & Exhibition Center is situated in western Haikou City, Hainan Province. Adjoining to the sea on the north side, its land is surrounded by hotels, apartments and a city park. The center provides exhibition and conference functions and the related ancillary facilities.

Its main part is so integrally designed that the exhibition center and conference center are combined to a huge and complete mass that, sitting besides the sea, becomes the terminal point and climax of the north-south landscape axis of the city. As the foreground of city axis, this huge building has a distinct axis that coincides with the center of the new urban area. It also includes many asymmetrical architectural shapes. All these make it a changing and rhythmically regular architecture with a coordinated overall shape. Its architectural image is an abstract manmade object appearing to be "sea water", "cloud", "beach", "marine creatures" and "ocean sight" that tally with the temperament of ocean. The steel structure of the regularly wavy shell roof consists of a combination of steel pipe trusses with equal cross-sections and is arranged in grids at the intermediary area and steel spatial structures at the edge area. The steel members are prefabricated at manufactories and fabricated and hoisted to place on site. They not only create unique building appearance but also highly unify with the interior space. Moreover, spaces for exhibition, conference, entertainment, and service of different scales and purposes are created. The louvers and high side lights of the shell roof units provide sufficient natural light of varying angles to the interior space. All around the exterior of the building are semi-arch eaves and canopies, as well as "stone benches" designed in line with windowsills.

The architecture has been built with stone, cambered aluminum plates, cast-in-situ GRC, and roof coating. Most of the interior steel structure is fully exposed to present its aesthetics and rhythm.

Photographs: Zhang Guangyuan, Li Xinggang, Zhang Yuting

俯瞰 / Aerial view

建筑造型的设计意向 / Reference of architecture form

屋顶模型研究 / Model study of roof

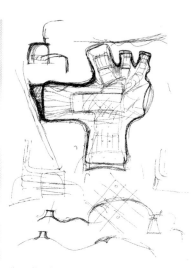
草图 / Sketch

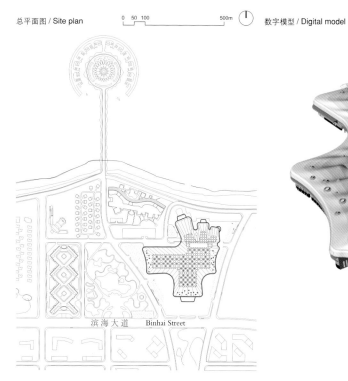
总平面图 / Site plan

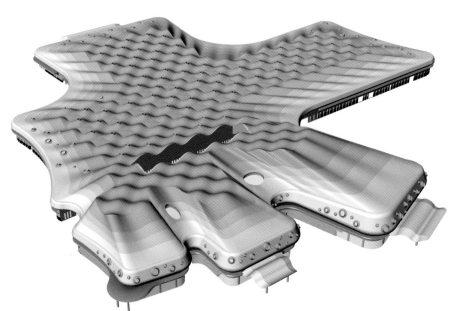
数字模型 / Digital model

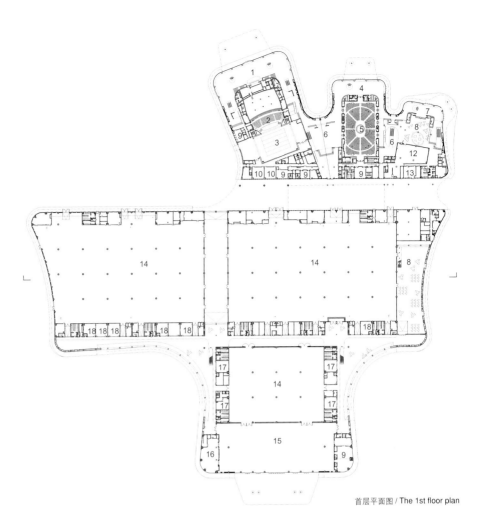

首层平面图 / The 1st floor plan

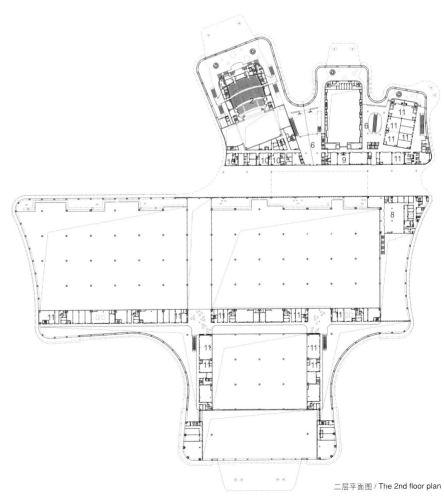

二层平面图 / The 2nd floor plan

1	剧场门厅	Antrance of theatre
2	观众厅	Auditorium
3	舞台	Stage
4	多功能厅门厅	Entrance of multi-function hall
5	多功能厅	Multi-function hall
6	共享大厅	Main hall
7	会议门厅	Entrance of conference centre
8	餐厅	Restaurant
9	贵宾休息厅	VIP room
10	化妆间	Make-up room
11	会议室	Meeting room
12	厨房	Kitchen
13	消防控制室	Fireproof control room
14	展厅	Exhibition hall
15	主门厅	Entrance of exhibition centre
16	公共接待	Reception
17	服务	Service
18	商店	Shop

东南俯瞰 / Southeast aerial view

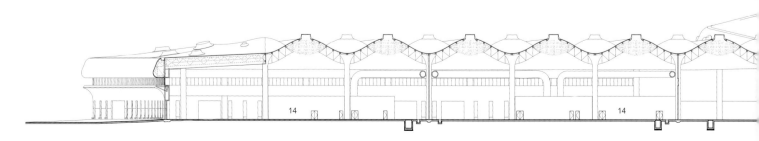

展厅室内 / Interior of exhibition hall

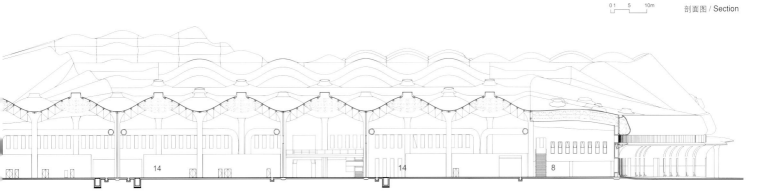

剖面图 / Section

125

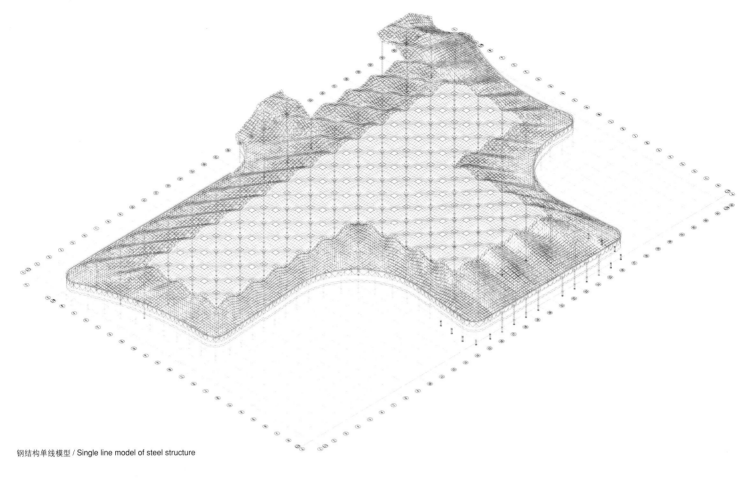

钢结构单线模型 / Single line model of steel structure

屋顶结构示意图 / Diagram of roof structure

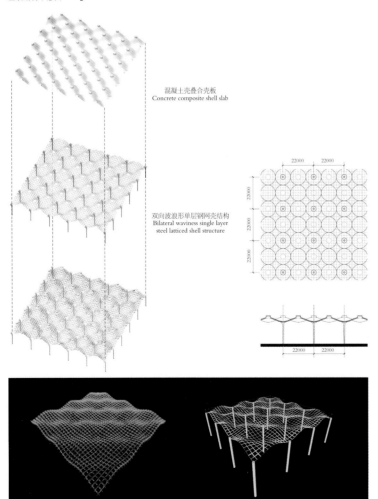

混凝土壳叠合壳板
Concrete composite shell slab

双向波浪形单层钢网壳结构
Bilateral waviness single layer
steel latticed shell structure

西南俯瞰 / Southwest aerial view

会议中心前广场 / Square in front of conference center

檐廊 / Veranda

展厅室内 / Interior of exhibition hall

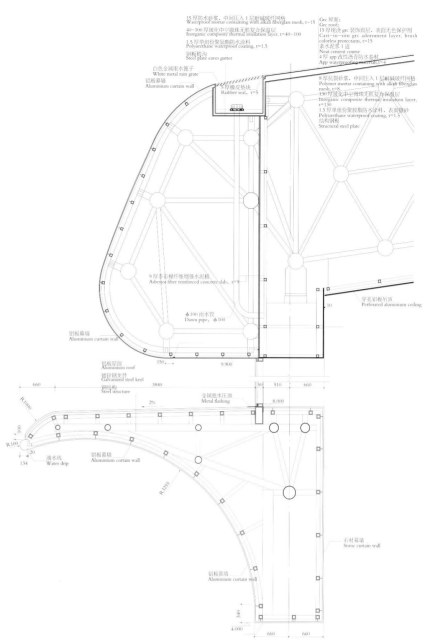

墙身详图 / Details of wall

施工中的展厅室内 / Interior of exhibition hall in construction

屋顶钢网壳结构 / Steel shell structure of roof

施工中的屋顶 / Roof in construction

天窗详图 / Details of louver

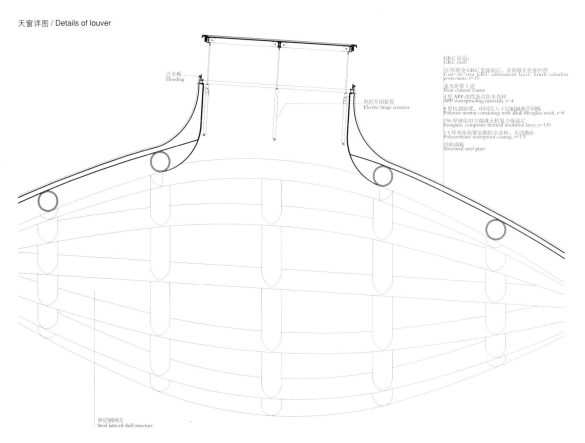

屋顶钢网壳结构施工 / Construction of steel shell structure of roof

模型研究 / Model study

绩溪博物馆 绩溪
Jixi Museum, Jixi

2009 至今

绩溪博物馆位于安徽绩溪县旧城北部，基址曾为县衙，后建为县政府大院，现因古城整体纳入保护修整规划，改变原有功能，改建为博物馆。包括展示空间、4D影院、观众服务、商铺、行政管理、库藏等功能，是一座中小型地方历史文化综合博物馆。

建筑设计基于对绩溪的地形环境、名称由来的考察和对徽派建筑与聚落的调查研究。整个建筑覆盖在一个连续的屋面之下，起伏的屋面轮廓和肌理仿佛绩溪周边山形水系，是"北有乳溪，与徽溪相去一里，并流离而复合，有如绩焉"的"绩溪之形"的充分演绎和展现。待周边区域修整"改徽"完成，古城风貌得以恢复后，建筑将与整个城市形态更加自然地融为一体。

为尽可能保留用地内的现状树木（特别是用地西北部一株700年树龄的古槐），建筑的整体布局中设置了多个庭院、天井和街巷，既营造出舒适宜人的室内外空间环境，也是徽派建筑空间布局的重释。建筑群落内沿着街巷设置有东西两条水圳，汇聚于主入口大庭院内的水面。建筑南侧设内向型的前广场——"明堂"，既有徽派民居的布局特征，也符合中国传统的"聚拢风水之气"的理念；主入口正对方位设置一组被抽象化的"假山"。围绕"明堂"、大门和水面有对市民开放的、立体的"观赏流线"，将游客引至建筑东南角的"观景台"，俯瞰建筑的屋面、庭院和秀美的远山。

规律性组合布置的三角屋架单元，其坡度源自当地建筑，并适应连续起伏的屋面形态；在适当采用当地传统建筑技术的同时，以灵活的方式使用砖、瓦等当地常见的建筑材料，并尝试使之呈现出当代感。

Jixi Museum is situated in the northern part of the old town of Jixi County, Anhui Province. The museum is built according to an integrated protection and rectification plan of the ancient town, including exhibition spaces, a 4D cinema, visitors' service, shopping, administration and storage functions.

The design has been based on a survey on the landscape, name origin survey of Jixi, and an investigation & study on the Huizhou-style settlements. The entire building is covered under a continuous roof that simulates the mountains and waters surrounding the county with its wavy profile and texture. It is the logical result and exhibition of "Jixi's shape". In Chinese, Jixi means the spinning of two streams, the Ru stream to the north and the Hui stream that is 1 li away, which first run in parallel to each other, then go further apart, and then re-merge. After the surrounding area is modified to the Huizhou style, the museum will be more naturally integrated with the entire city configuration.

To maximally preserve the existing trees on the land, the overall layout of the building consists of multiple courtyards, patios and lanes to create comfortable spaces and reinterpret the spatial layout of the Huizhou-style architecture. Two streams along the lanes, one in the east and the other in the west of the cluster of buildings, merge to the pool in the large courtyard at the main entrance. In the southern part of the building is "Ming Tang", which is an enclosed front square in line with the layout of Huizhou-style houses and the traditional Chinese Fengshui principles (Chinese geomancy). Right opposite to the main entrance are a set of abstract "artificial rockworks". Surrounding "Ming Tang" is a stereoscopic "sightseeing route" that guides tourists to the "sightseeing platform" at the southeast corner of the building, where they can command the roof-scape.

Regularly arranged collar roof units adopt a slope originated from local buildings and are adapted to the wavy and continuous roof shape. Common local building materials such as bricks and tiles are used in a flexible manner and trialed to have a contemporary appearance.

摄影：李兴钢

Photographs: Li Xinggang

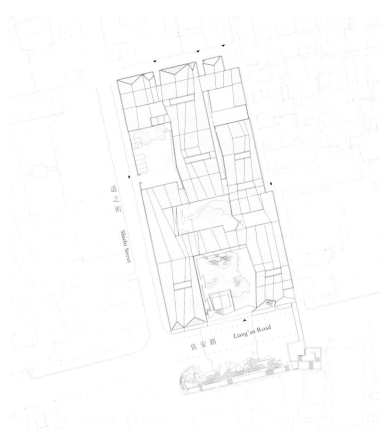

总平面图 / Site plan

模型研究 / Model study

1	庭院／天井	Courtyard / Patio
2	序言厅	Preface hall
3	接待厅	Reception hall
4	贵宾厅	VIP room
5	教室	Classroom
6	商店	Shop
7	售票	Ticket
8	茶亭	Tea pavilion
9	保留县衙遗址	Reserved relic site
10	展厅	Exhibition hall
11	4D 影院	4D cinema
12	临时展厅	Temporary exhibition hall
13	报告厅	Lecture hall
14	设备用房	Equipments
15	消防控制室	Fireproof control room
16	技术和管理用房	Technology and management room
17	临时储藏	Temporary storage
18	藏品设施空间	Storage facilities
19	街巷	Lane

首层平面图 / The 1st floor plan

施工中的展厅 / Exhibition hall in construction

施工中的接待厅 / Reception hall in construction

A–A 剖面图 / Section A–A

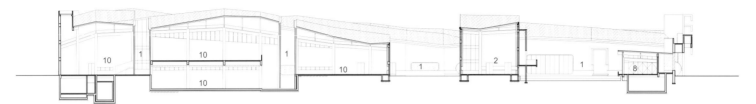

B–B 剖面图 / Section B–B

南立面图 / South elevation

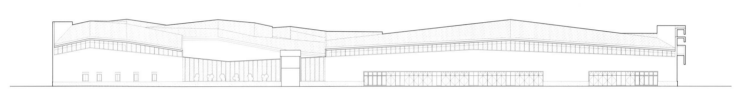

西立面图 / West elevation

墙身及屋顶详图 / Details of wall & roof
施工现场俯瞰 / Construction site aerial view

小青瓦屋面（钢屋架）
Cyan tiles roof (steel roof truss)
小青瓦用20厚1:1.4水泥石灰砂浆加水泥重的3%麻刀卧铺
Cyan tiles thick with 1:1.4 cement mortar t=20, with 3% MaDao berth
1.5厚聚氨酯防水涂膜
Polyurethane waterproof coating t=1.5
3厚APP防水卷材
APP waterproofing materials t=3
最薄处30厚C15细石混凝土，内铺钢丝网
C15 fine stone concrete t=30, with steel wire inside
波形钢板，波高35，双向搭接不小于50mm
Corrugated steel sheet, wave height=35, two-way lap are not less than 50 mm
140高槽钢檩条，内填60厚挤塑板保温层
U-steel purline h=140, with plastic extruded board thermal insulation layer t=60
9厚防火石膏板两层用自攻螺丝固定
Fire-proof gypsum board t=9 two layer with tapping screws fixed
满刷乳化光油防潮涂料2道
Full brush light oil emulsion paint 2 times
满刮2厚面层耐水腻子
Full blow surface water resistant putty t=2
白色涂料饰面
White paint facing

高度控制基准点
Height control starting point

水泥钉@900，下垫镀锌垫片
Cement nail@900, galvanized gaskets

2厚防火石膏板两层，表面白色喷涂
Fire-proof gypsum board two layer, white paint

槽钢檩条
U-steel purline

40厚挤塑保温层
Plastic extruded board thermal insulation layer t=40

钢屋架，表面白色喷涂
Steel roof truss, white paint facing

200x200x16预埋件 5φ12
Embedded part 200x200x16, 5φ12

小青瓦 180×150 鱼鳞纹拼花
Cyan tiles 180×150 fishscale pattern

电动通风排烟窗
Electric ventilated exhaust window

方钢100x150x6，深灰色
Deep gray square steel 100×150×6

轻钢龙骨石膏板内墙，表面白色喷涂
Light steel keel plasterboard interior wall, white paint facing

泄水孔
Weep hole

双层百叶通风口
Double shutter vents

湿贴石材
Stone wet combining

空调机
Air conditioner

白色涂料外墙（采用传统工艺）
White paint exterior wall (the traditional craft)
25厚白色涂料饰面
White paint facing t=25
搓3-5厚聚合物砂浆中夹耐碱玻璃纤维网格
Polymer mortar containing with alkali fiberglass mesh t=3-5
聚合物砂浆贴30厚挤塑板保温层
Polymer mortar stuck plastic extruded board thermal insulation layer t=30
20厚聚合物砂浆找平层
Polymer mortar screed-coat t=20
外墙墙体
Exterior wall

地面作法详装修设计
Detail by decoration design

落水链，304不锈钢
Reinforced concrete roof slab

密封膏
Sealant

条石
Stone slab

湿贴石材
Stone wet combining

4 φ 16
8@200

石膏板吊顶
Plasterboard condole top
现浇钢筋混凝土板预留φ10钢筋吊环，中距横向≤1200，纵向1100
Cast-in-place reinforced concrete board φ=10 reserve reinforced rings, d ≤ 1200 in horizontal, 1100 in vertical
U型轻钢龙骨 CB60×27 中距429,10号低碳镀锌钢丝吊杆，与预留吊环固定
U steel keel CB60×27 d=429, No.10 low carbon galvanizing steel wire for the boom, obligate rings fixed
9.5厚纸面石膏板，用自攻螺丝固定
Paper face plasterboard t=9.5, with tapping screws
满刷乳化光油防潮涂料2道
Emulsifying oil moistureproof light brush paint 2 times
U型轻钢龙骨 CB50×20 中距1200
U steel keel CB50 × 20 d=1200
满刮2厚面层耐水腻子
Full blow surface water resistant putty t=2
白色乳胶漆
White emulsioni paint

1 φ 8
φ8@200

湿贴石材
Reinforced concrete roof slab

2 φ 6
φ 6@200

内墙涂料墙面
Interior wall of painting
白色合成树脂乳液涂料
White synthetic resin emulsion coatings
2厚精品粉刷石膏罩面
High-quality plaster coverface paint t=2
5厚粉刷石膏砂浆打底分遍刮平
Paint the plaster mortar make up t=5
3厚外加剂专用砂浆抹基面粗糙
Admixtures with special mortar base surface roughness scraping t=3
聚合物水泥砂浆修补墙面
The polymer cement mortar repair wall

100厚石材勒脚，密缝砌筑
Stone plinth close joint t=100

石材沟盖板
Stone trench cover

预埋铁件
Embedded iron piece

C30 混凝土 φ 8@150
C30 concrete φ 8@150

C15 混凝土
C15 concrete

远景 / Distant view

元上都遗址工作站　正蓝旗

Entrance for Site of XANADU, Zhenglanqi

2010 ~ 2011

　　元上都遗址工作站位于元上都遗址之南，解决了遗址景区售票、警卫监控、管理办公、休息及游客公共卫生间等功能需求，并配合元上都遗址申报世界文化遗产。基地选址于景区现状入口处，将题有"元上都遗址"的原有门楣和刻有元上都遗址地图的石碑设置于正对遗址轴线的延长线上，而将新建的建筑、原有的"忽必烈"雕塑和电瓶车停车场等偏于轴线东侧布置，以留出面向遗址的景观视线通廊。

　　一组白色坡顶的圆形和椭圆形小建筑，围合成对内和对外的两个庭院，分别供工作人员和游客使用。根据功能需求，这些小建筑大小不一、高低错落，相互之间的群体关系形成了有趣的对话。

　　圆形和椭圆形的建筑形体朝向庭院的部分，在几何体上连续地切削，形成像建筑被剖开后展开的折线形内界面，采用清水混凝土做法（后覆上一层薄薄的白色涂料）；建筑形体朝向外侧的连续弧形界面，则罩以白色半透明的PTFE膜材，引发蒙古包的联想，带来草原上临时建筑的感觉，最大限度降低对遗址环境的干扰。膜与外墙之间空隙里隐藏的灯管将在夜晚发出白色的微光，更显轻盈，似乎随时可以迁走一样，暗合草原的游牧特质，同时表达了对遗址的尊重。

Entrance for Site of XANADU is situated to the south of the ruin of the Yuan Dynasty Upper Capital (XANADU), providing auxiliary functions such as ticketing, guarding, management offices, resting, and tourists' lavatories for the scenic area of the ruin. The entrance is located at the spot of existing entrance, and includes an existing door head bearing inscription of "元上都遗址" (Site of XANADU) and a piece of stone bearing inscription of a map of the Site along the extension line of the axis. The new buildings, as well as the existing "Kublai Khan" statue and battery car parking, are located to the east of the axis, leaving a visual passage to the landscape of the Site.

A group of white circular and oval buildings with pitched roofs connected with each other enclose a courtyard for staff and another one for tourists. Depending on the function, these small buildings have varied sizes and heights, and their relationship between each other as part of a group creates an interesting dialog.

The side of the circular and oval buildings is continuously "cut" to form a polyline-enclosed interface as if the volume were cut and unfolded. This concrete interface has an as-cast finish covered by a thin sheet of white coating. Facing outwards, the buildings have a continuous arc-shaped interface covered by white translucent PTFE membrane to remind visitors of Mongolian yurts. Thus, the buildings seem to be temporary. In the clearance between the membrane and the exterior wall are hidden lamp tubes that emit faint white at night, making the buildings more light-footed as if they could be moved away to suit the nomadic living and express a respect to the Site.

摄影：张广源、邱涧冰、李宁、赵小雨

Photographs: Zhang Guangyuan, Qiu Jianbing, Li Ning, Zhao Xiaoyu

全景 / Panoramic view

模型研究 / Model study

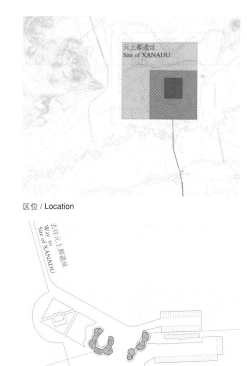

区位 / Location

总平面图 / Site plan　　0 5　10　　30m

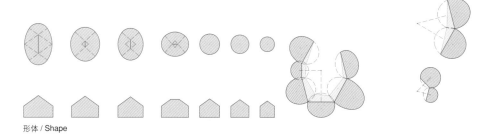

形体 / Shape

灯光模型研究 / Model study of lighting

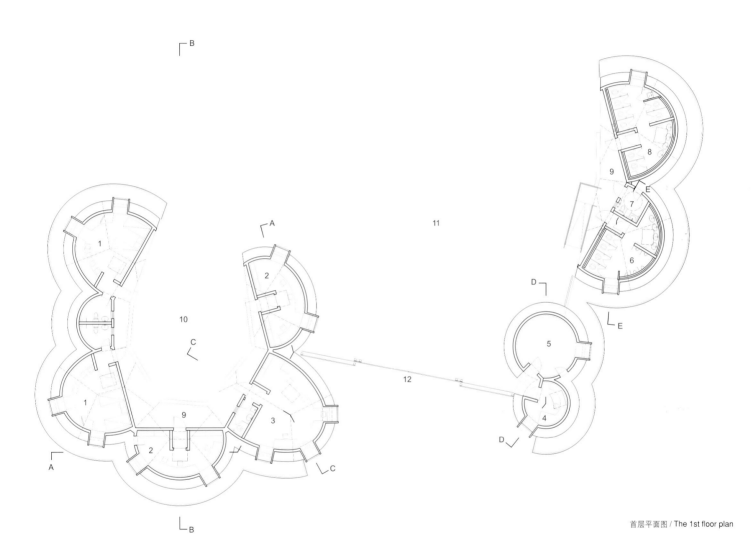

首层平面图 / The 1st floor plan

1	宿舍	Dormitory
2	办公室	Office
3	警卫监控室	Security control room
4	售票处	Ticket office
5	储藏间	Storage
6	男卫生间	Male lavatory
7	无障碍卫生间	Barrier-free lavatory
8	女卫生间	Female lavatory
9	檐廊	Veranda
10	庭院	Courtyard
11	广场	Square
12	大门	Gate

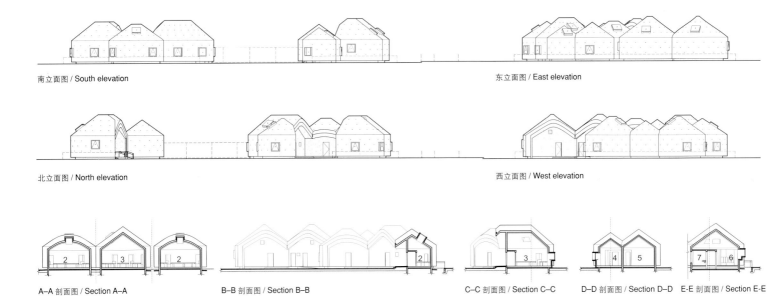

南立面图 / South elevation

东立面图 / East elevation

北立面图 / North elevation

西立面图 / West elevation

A–A 剖面图 / Section A–A

B–B 剖面图 / Section B–B

C–C 剖面图 / Section C–C

D–D 剖面图 / Section D–D

E-E 剖面图 / Section E-E

檐口展开图 / Eave spreading drawing

檐下墙体展开图 / Wall spreading drawing

模型研究 / Model study

屋顶局部 / Roof details

近景 / Close view

庭院 / Courtyard

全景 / Panoramic view

大草原上的小房子 / The small buildings on the prairie

147

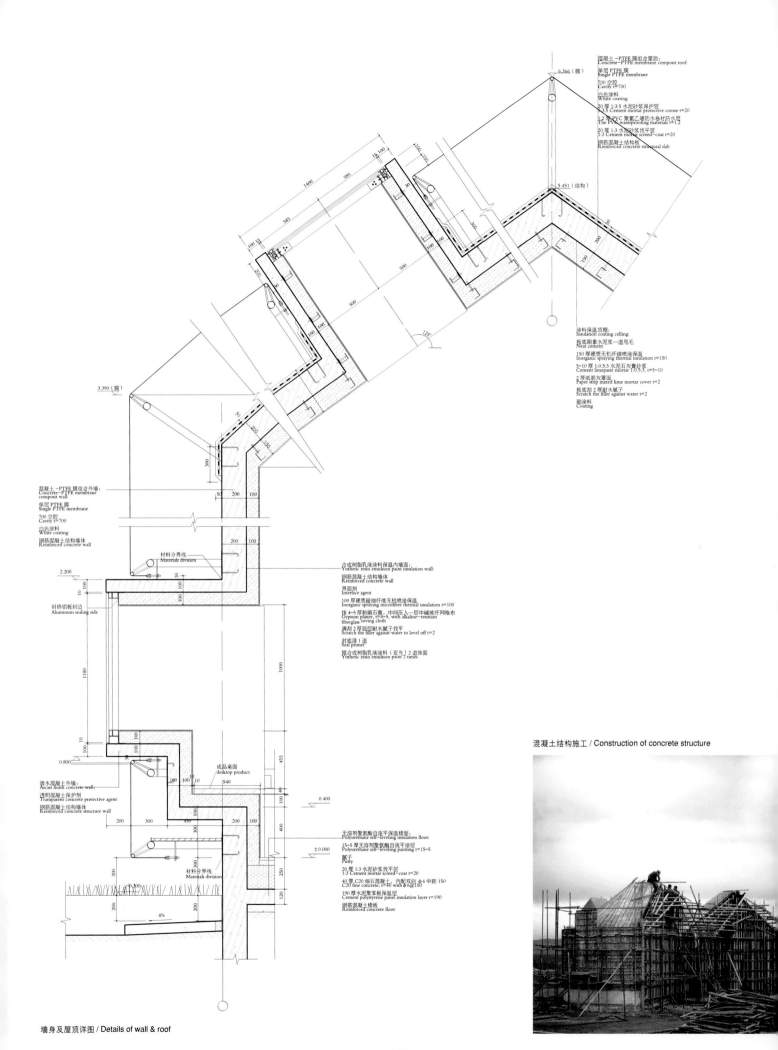

墙身及屋顶详图 / Details of wall & roof

混凝土结构施工 / Construction of concrete structure

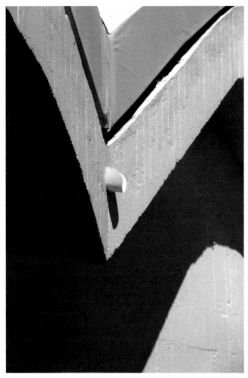
檐口细部 / Eaves detail

外窗细部 / Window detail

办公室室内 / Interior of office

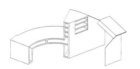
售票处家具 / Ticket office furniture

办公室家具 / Office furniture

警卫监控室家具 / Security control room furniture

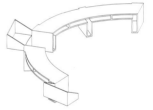
宿舍家具 / Dormitory furniture

膜结构施工 / Construction of membrane structure

以保留树木为中心的组团空间 / Public space around retained tree

西柏坡华润希望小镇　　西柏坡
Xibaipo China Resources Hope Town, Xibaipo
2010 至今

西柏坡华润希望小镇位于河北省平山县西柏坡镇霍家沟，是华润集团捐资兴建的新农村示范项目。项目由原有的三个相邻的山村集中组合而成。包括238户农宅和村委会、村民之家、卫生所、幼儿园、商店、餐厅等公共设施。基址位于山坳的低洼处，三面环山，一面朝向水库。设计保留了基址中现存的泄洪沟并进行适度修整，结合新设置的村民广场及公共设施，构成贯穿整个小镇的景观带和公共活动中心，并顺势将小镇分为三个居住组团。建筑主要依坡地而建，低洼部分的基址则垫高为可用于房屋建设的台地。公共设施位于小镇中心并在入口处标示出小镇的边界。由此，形成了"居于高台，游于绿谷，聚于中心"的立体化聚落空间。

设计中尽可能地保留了场地中的树木和一户质量尚好的农宅院落以及一座古桥，作为新聚落的历史记忆。以多样的方式处理建筑群组、公共节点、道路桥梁和街巷院落，营造丰富多变的路径、视角、景观以及停驻、交往、活动空间。农宅的设计在研究当地传统民居空间特征的基础上，以L形建筑主体加院墙形成可重复的围合院落，并对风水传统、朝向、檐下空间、自然通风、个性化、私密性、未来加建以及"农家乐"主题旅游功能进行了充分的考虑。

华润希望小镇的设计运用了聚落的设计方法，从宏观、中观以至微观的不同层次应对场地条件、人群特征和传统沿承的矛盾，因地制宜，从中自然生成的秩序使其具有了聚落的独特性和丰富性。

Xibaipo China Resources Hope Town is situated at Xibaipo Town, Pingshan County, Hebei Province. The Hope Town is New Village Demonstration Project invested by China Resources Group. It includes 238 farmhouses supported by a village government building, "villagers' family", clinic, kindergarten, shops, restaurants, and other public facilities. The town sits on a selected low-lying lot surrounded by mountains and a water reservoir on the other side. The design reconditions an existing waterway which can discharge flood from mountains out of the site, and use it as a public landscape and activity zone running through the entire town. Farmhouses are mainly built along the hill-slope area, and the low-lying area is filled to a higher elevation to form a platform for houses. Public buildings are located in the center of town and mark the boundary of the town at the entrance. The above results in a stereoscopic settling space where "villagers live on a high platform surrounded by green valley and gather at the center".

Trees, a well-preserved courtyard farmhouse, and an ancient bridge are retained reminding the history of the new settlement. In addition, the building groups, public nodes, roads & bridges, streets & lanes, and courtyards are differently treated to create a variety of routes, view angles, landscapes, and spaces for villagers to communicate. The farmhouses are designed based on the spatial characteristics of traditional local place. Each farmhouse consists of an L-shaped main building and a repeatable enclosed courtyard. The design also considers the Fengshui tradition, orientation, natural ventilation, personalization, privacy, later expansion, and possibility for "rural house" tourism.

The town has been designed by method of settlements to address the conflicts in site conditions, population characteristics, and continuity of tradition on the macro, medium, and micro levels, and to suit the measures to differing situations of locality, which gives birth to a unique settlement with abundant features.

摄影：张广源、李兴钢、邱涧冰、梁旭、马津

Photographs: Zhang Guangyuan, Li Xinggang, Qiu Jianbing, Liang Xu, Ma Jin

小镇远眺 / Distant view of town

模型研究 / Model study

总平面图 / Site plan

岗南水库 / Gangnan Reservoir

西柏坡公路 / Xibaipo Expressway

组团空间俯瞰 / Aerial view of public space

公共建筑首层平面图 / The 1st floor plan of public buildings

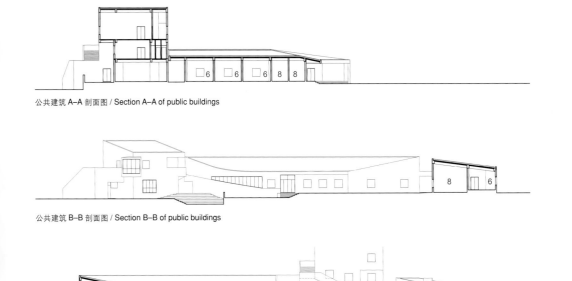

公共建筑 A–A 剖面图 / Section A–A of public buildings

公共建筑 B–B 剖面图 / Section B–B of public buildings

公共建筑 C–C 剖面图 / Section C–C of public buildings

1	幼儿园	Kindergarten
2	村民之家	Villagers family
3	卫生所	Clinic
4	村委会	Village committee
5	餐厅	Restaurant
6	商店	Shop
7	农资站	Department of agricultural resource
8	服务	Service
9	厨房	Kitchen
10	绿谷	Green valley

公共建筑外景 / Exterior of public buildings

公共建筑夜景 / Night view of public buildings

小镇俯瞰 / Aerial view of town

住宅外景 / Exterior of houses

花墙砌法研究 / Study of leacky-wall bricklaying

A210 单元首层平面图 / The 1st floor plan of unit A210

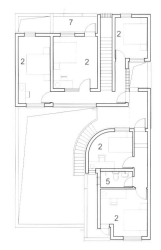

A210 单元二层平面图 / The 2nd floor plan of unit A210

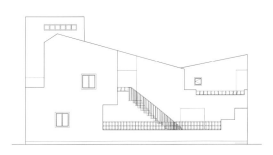

A210 单元西立面图 / West elevation of unit A210

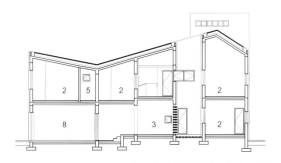

A210 单元剖面图 / Section of unit A210

1	起居室	Living room
2	卧室	Bedroom
3	餐厅	Dining
4	厨房	Kitchen
5	卫生间	Lavatory
6	储藏	Storage
7	阳台	Balcony
8	车库	Garage

李兴钢访谈

采访人／黄元炤
北京　2011.10.18

黄：就我的观察，您设计上有两条路线。一条路线是关注到中国古典园林中的一种转折的指向性，如：兴涛接待展示中心；另外一条是因为参与"鸟巢"设计，开始关注到表皮，倾向于一种皮层－造型的设计语言，比如说像复兴路乙59-1号改造等。

先来说说前者。2002年，兴涛接待展示中心，是您初期建筑实践中一项重要的代表性作品，似乎看到您内心潜藏的设计追求。给我最深的印象就是白色的墙板，一下转折成为垂直的墙，一下又转折回水平的板，形成包覆在建筑外部的骨架，并界定出内外空间。我觉得转折连续的墙体，有一种指向性，以墙体的实，围塑出虚的空间，且暗示人的视觉与移动，有点中国古典园林中步移景异的意味，您利用墙体的水平与垂直的转折连续，创造出一个视觉出口或是路径，从一个场景转折到另一个场景，往上往下、往左往右不停地转换，我的观察，您用一种全新的透视性与延展性的构造物，去推翻一般接待中心或枯燥僵化或过度喧闹的印象，这部分您有何看法？

李：这个项目对我有特别的意义，它是800多平方米的一个小项目，所以我可以做得相对轻松一些。另外，它比较体现我的独立性的思考，然后也把它表达得比较干净和纯粹一些。

黄：就是允许您在一个范围内可以去做得相对来说条件更开放。

李：对，而且我是希望能够把我做的东西，做得跟别人不一样一些。我觉得即使是面对一个住宅小区的配套学校或者一个售楼处，也可以有机会表达自己的思考，这是我当时的一个出发点。

你刚才的观察很对，我实际上始终对建筑里面的中国性这样一个课题感兴趣。兴涛学校的设计主题实际上是延续我的本科毕业设计里已经在思考的中国城市和建筑之间的关系。兴涛接待展示中心则开始反映我对中国园林的观察和兴趣。而为什么做得轻松呢，那时我并没有把古典园林都研究透了，理论体系是怎么，空间体验是怎么，当然我现在在思考的更多，但那个时候，我觉得当时就是那样一个片段性的感觉，对园林的，把这个感觉表达出来就是了。

我觉得墙对园林空间、景观、人的感受的引导特别有意思，墙至少是园林里最重要的要素之一吧。这个小小的接待中心，里面有若干不同的功能，

Interview

By Huang Yuanzhao
Beijing 2011.10.18

Huang: I have observed that you follow two design routes. The first focuses on the directivity of transition in classical Chinese gardens as in Xingtao Reception & Exhibition Center. The second began from your participation in the "Bird's Nest" design. It focuses on the skin, that is, you tend to adopt a skin-shape design language in, for example, Reconstruction of No.B-59-1, Fuxing Road.

Now let's see the first route. Xingtao Reception & Exhibition Center designed in 2002 was one of the important representative works in your early practice, which exposed the design pursuit hidden in your mind. I am most impressed by the white wall-slabs, transiting suddenly from vertical to horizontal and forming an external skeleton. The physical wall encloses a virtual space and guides the vision and movement of people. This somewhat likes classical Chinese gardens, where the scene changes with the position of visitors. When the wall horizontally and vertically transits but remains continuous, a visual outlet or path was created. The scene continuously changes from one to another, either up and down or left and right. I think you were using a structure with a brand new perspective feature and extensibility to remove the dull or noisy impression of normal reception center buildings. So what do you say about it?

Li: This project was especially significant to me. Since it was a small project of 800m², I did it relatively easily. In addition, it reflected my independent thinking, which was expressed clearly and purely.

Huang: That is, you were allowed to do it more freely to some extent.

Li: Right. I hoped to do something special and express my thought. This was one of my starting points at that time.

What you have observed is exactly right. I'm always interested in Chinese features in architecture. Xingtao Exhibition & Reception Center indeed reflected my first observation and interest in Chinese gardens. I did it easily because I hadn't clearly understood classical gardens, theoretical system and spatial experience then. The design was just an expression of fragmental feel of gardens. Certainly I'm thinking more right now.

I was interested in how walls could guide the space and landscape of gardens and human feeling. Walls were at least one of the most important elements in garden. Despite the miniature size, the reception center provided different functions. For example, it was a symbolic gate to the residential quarter and contained an exhibition area for display boards and models. In addition, there was also a sample unit for visiting. All these functions had internal connection with one another but different spatial

比如它是小区的一个象征性的大门，然后有展板、模型的展示区，然后因为小区已经部分建成，要看实景，另外还有一个参观的样板间，得把这些功能合理地串起来，它们有内在的联系，但它们又有不同的空间要求和视线需求，我觉得对人来讲，其实这有点像"游玩"——游览和玩赏，我可以把它做成一种游园的方式和行为。

黄：人随墙走，墙随人动。

李：对，就是这个感觉——游园。而引导者是墙，垂直的墙体，往上水平延伸就变成楼板，用墙和板的这样一种延伸，把整个建筑的流线和空间给带起来。

黄：所以说这个项目，较之以往的项目，您可以更出来一点，可以更放松地表述自己所想要的设计思考。

李：对。我当时还试着申报了一个英国的世界建筑奖，并获得了提名奖，应该算是中国建筑师获得的比较早的国际奖项。我其实就是想去试一下，这样一个作品在国际平台上，人家是怎么来判断的，所以当时我觉得还是挺高兴的，虽然只是一个提名奖，但可能给自己带来一些自信。

黄：就我的了解，北京的西环广场暨西直门交通枢纽，是您与法国AREP公司的合作项目，这个项目给我的感觉，就是想创造出城市中鲜明的地标建筑，采用无标准层设计，自下而上，面积从约900平方米至2000平方米逐层递增，这是一个重点；而另一个重点是，外立面为曲线型设计，结合纯净的玻璃幕墙，几道流畅的弧线勾勒出外形，与一般商务建筑群不同，它突破了现代写字楼直角平面式的工业化造型，能谈谈这个合作项目吗？您在这里面参与的程度多大？

李：我觉得它应该说是一个典型的工程性项目，我在法国学习的时候，实习是在法铁的火车站研究局，他们有一个对外的公司AREP，很擅长交通建筑的设计，我们就邀请他们一块儿参加西直门交通枢纽的竞赛。竞赛的整个构思阶段，是中法建筑师一起进行的，中方这边崔总（崔愷）为主，我是中方的第二把手。方案构思是合作团队的共同成果，赢了竞赛之后的工程设计阶段则是由我担任设计主持人，当然也有法方的很多参与，特别是初步设计。

requirements and visual needs. In my opinion, they were something like those to be toured and appreciated by visitors. I could make them being accessed in a manner of garden tour.

Huang: The walls seem to move when visitors walking along.

Li: Right. It was the feel of garden tour guided by the wall. When the vertical wall extended upwards, it became a horizontal floor slab. As the wall and slab extended, they led the flowing line and space of the entire building.

Huang: So you could more easily and freely express the design thoughts in this project than in any previous one.

Li: That's it. I sent it to compete for the World Architecture Awards in UK and won a finalist prize, which should be one of the early awards that Chinese architects had achieved. I did this just to see how such a work would be judged on an international platform. Therefore, I felt very happy about the prize, which, although just a nominee, might make me more confident.

Huang: As I know, Xihuan Plaza & Xizhimen Transportation Hub was your joint project with the French architectural firm AREP. I feel that the project was to create a landmark. One of its highlights is the building is designed without any standard floor and the floor area decreases from about 2,000m^2 to 900m^2 from the bottom up. Another highlight is the curved external facade that combines with clear glass curtain wall. The fluent arc lines define the outline that is different from any industrial design of modern office buildings that consist of right angles and planes. How much were you involved in it?

Li: I think it should be a typical engineering project. When I studied in France, I interned at Railway Station Research Bureau of SNCF. AREP is its subsidiary and is good at designing transport architecture. So we asked AREP to work with us to compete for Xizhimen Transportation Hub. The entire conception process was jointly carried out by French and Chinese architects. The Chinese team was led by Cui Kai, Chief Architect of China Architecture Design & Research Group, and I was the deputy leader. The conceptual design that won the competition was the joint achievement. In the engineering design stage that followed, I was the superintendent. Certainly the French architects were much involved, especially in the preliminary design.

Xihuan Plaza was designed as three towers rather than a single giant. One reason was to keep the visual passage from the northwestern part of the 2nd Ring Road to the tip of the tower of Beijing Exhibition Theatre, although this hadn't been required in the urban planning. The other reason was for

西环广场设计为三座塔楼而不是一座集中的庞大楼体，一个原因是想保留原来从西北二环路到北展剧场尖塔顶的视觉通廊，虽然这一点在城市规划上并没有要求；另一个原因是业主也会喜欢，因为三栋相对独立的写字楼更有利于销售。

之所以把塔楼做成曲线的形式，与原来曾经在这里存在过的西直门城楼的门洞有那么一点隐含的联想之意，算是对此处那个古老城市之门的某种记忆的唤起。这个项目非常复杂，一个是中外的合作设计模式，一个是项目本身的功能复合性：有地铁、城铁、国铁等多种交通模式要结合在一起成为交通换乘枢纽，另外还有各种类型的商业、写字楼等，即使放在今天，都是一个很复杂的大型城市综合体，何况是十多年前。对我的一个比较大的收获，是对大型工程的历练、控制和主持，拥有技术和团队的把控能力和自信心，是很重要的锻炼，非常有好处的。

所以，西环广场暨西直门交通枢纽实际上是对于后面主持"鸟巢"的工作一个非常好的铺垫；而且还有一个收获，是中外合作设计的模式，怎么样能找到一种更好的、更平等的方式。因为在这之前，国内外建筑师的合作，通常是老外先做好方案，国内再做施工图；而由西直门这个项目开始，我们开创了这样的合作模式：从竞赛和概念设计阶段就在一起平等地探讨、画草图，然后共同决定设计方向；在工程设计阶段也是基本全程地共同工作。

黄：算是一个对等关系，中外对等关系。

李：至少是比以前更对等吧，就是说相互都有学习借鉴，从方案、初步设计到施工图，针对不同的阶段、不同的情况和条件，相互派人到对方的办公室工作，互相交流、互相促进。我觉得经由这些交流，帮助我有一种跟外国建筑师打交道很自如、很自信的一个状态，同时也有助于必要时候思维方式的国际化的转变，至少不因文化的差异而产生太大的隔阂和障碍。我觉得这种状态还是挺重要的。

黄：2003年，您去了瑞士，参与北京奥运会主会场——中国国家体育场的国际竞赛，担任中方的设计负责人。就我的观察，"鸟巢"是由不规则钢结构编织而成的椭圆马鞍形，彷佛是从瓷器的古雅意韵中衍生出的外形，体现了中国传统文化中特有的智慧与魅力。而就建筑学的观点，它表现出来

the preference of the owner because three office buildings independent to each other could be more easily for sale.

The towers were made curved to imply the opening of Xizhimen Gate Tower once existed there, thereby reminding viewers of that ancient gate to the city. The project was rather complicated because of its design mode of Sino-France cooperation and its versatile functions including a transfer hub that combined subway, light rail, and railway. Furthermore, various business and office buildings were also required. It is a complicated urban complex even today, let alone it was done more than a decade ago. My major harvest was the experience, control, and direction of a large work like this, as well as the ability and confidence to control technologies and a team. The project was an important and beneficial practice.

So this project well prepared me for the later direction in the "Bird's Nest" work.

Another achievement was on the mode of better and more equal manner of cooperation with foreigners. Before this project, the cooperation had been carried out in such a way that foreign architects designed a plan and then the Chinese architects prepared the construction drawings. Beginning from the Xizhimen project, we created a cooperation mode in which Chinese and foreign architects equally and cooperatively discuss with each other, plot sketches, and determine a direction of design as early as from the competition and conceptual design stage to engineering design throughout the whole process.

Huang: This is an equal relationship.

Li: At least it is more equal than before. Both parties can learn from each other from the conceptual design, through preliminary design, to construction drawings. They may send persons to the other party's office for communication and assistance depending on the stage-specific situation and condition. I felt that the communication helped me freely and confidentially deal with foreign architects and think globally whenever necessary. At least, the barrier resulting from cultural difference was no more significant.

Huang: In 2003, you went to Switzerland to participate in the international competition of National Stadium, the main stadium of Beijing Olympics and you were the Chinese design person-in-charge. As I see, the elliptical-saddle shape of "Bird's Nest" was woven with irregular steel structures, as if it were derived from quaint ceramics that embodies the unique wisdom and charm of Chinese traditional culture. In view of architectonics, the building represented an architecture language of skin that was the membrane between the internal building and the external environment, or a filter between manmade work and

的是一个皮层的建筑语言，而这个皮层是建筑与环境内外之间的隔膜，是人工与自然之间的一个过滤器。您怎么看待"鸟巢"对您设计思想上的影响？还是您参与到这个项目后，得到了某种设计启发？

李：参与这个项目对我最重要的影响，其实是赫尔佐格与德梅隆的工作方法——他们是怎么来做建筑的。我觉得对于赫尔佐格与德梅隆来讲，他们把建筑师当成一个有点像工匠一样的行当，那么你做这个行当，怎么把这个事做出来，他们是有自己的一套做法的。他们的工作受到很多艺术家的影响，但是这种影响如何被转化成他怎么来做，他们又是怎么把一个设计从开始就控制好直到最后完成，这个过程我经历得非常完整和全面，并对我有很大的影响。

2003 年"鸟巢"中标后，我的工作室也同时成立，我就开始在自己的团队中建立这样一种类似的工作模式、工作方法。

黄：我们现在正好来说一下另外一条设计路线。复兴路乙 59-1 号改造与北京地铁四号线出入口似乎与"鸟巢"类似，也是由表皮演化出来的造型体现，强调的是多面向与多角度的皮层整体感，追求动态自由化的极致表述。而这个表皮其实我们都了解，在中国境内，大概是从"鸟巢"开始成为了中国当代的流行建筑语言，而您因"鸟巢"和赫尔佐格与德梅隆的接触及合作，这是否影响您日后的设计思维与操作，能说说您的看法吗？

李：我曾经写过一篇文章，是关于复兴路乙 59-1 号改造这个项目的，发表在建筑学报上。我的观点是复兴路乙 59-1 号不是一个简单化的、专门在表层的表皮上做文章的那种"表皮建筑"，虽然它外观所呈现的感觉像是那样。但我说即使一定要说它是表皮的话，也是一个空间化的表皮，并且它有自己内在的逻辑。

我觉得在这个建筑的设计上我受到的赫尔佐格与德梅隆的影响仍然是工作方法上的影响：怎样思考一个建筑的发生——线索是如何产生的，然后线索是怎样导致相应的手法，最后建筑的面貌如何决定和产生。我觉得如果说我受到他们影响的，就是在这个地方，而不是说是某种具体的建筑语言，比如有人说复兴路乙 59-1 号与北京地铁四号线出入口像"鸟巢"之类的，其实完全不是，甚至我还会刻意回避这样一种肤浅的相似。

the nature. How did "Bird's Nest" affect your design idea? Did you get some hint on your design after participating in the project?

Li: The working method of Herzog and de Meuron (HdeM), that was, how they designed the building, most significantly affected me when I was involved in this project. In my opinion, HdeM consider that an architect works in a trade that is somewhat like that of a craftsman. They had been influenced by many artists, and I was much affected by how the influence had been converted to their practice. Moreover, I experienced a complete process whereby they had well controlled the design from the beginning to the end.

My atelier was established when we won the competition of "Bird's Nest" in 2003. Then I began to setup a similar working mode and approach in my own team.

Huang: Well, let's talk about the other design route. Reconstruction of No.B-59-1, Fuxing Road and the Accesses of Line 4 of Beijing Subway seem to be similar to "Bird's Nest", because they are represented by shapes that come from skin, emphasize the skin integration of multiple directions and angles, and pursue ultimate representation of dynamics and freedom. However, we all know the skin, which might become a popular architecture language in contemporary China since "Bird's Nest". I wonder whether your contact and cooperation with HdeM in "Bird's Nest" have affected your design thinking and operation afterwards. What's your point?

Li: I published an article about Reconstruction of No.B-59-1, Fuxing Road on the Architectural Journal. In my opinion, the Fuxing Road building is not a plain "skin building" that has nothing worth mentioning other than its skin, although it seems like that. Even if it were skinny, it would be a spatially skinny building with the internal logic of its own.

I do think that HdeM's influence on my architectural design is still on the working method; that is, how to think about the occurrence of a building-how a trait is generated, how it results in the appropriate techniques, and how to decide and generate the architectural visage. Their influence on me is rather than a specific architectural language. For example, if somebody says that the Fuxing Road project and the Accesses of Line 4 of Beijing Subway are similar to "Bird's Nest", they are wrong. Moreover, I deliberately avoided such superficial similarity.

The irregular grids of the curtain wall of the Fuxing Road project were naturally generated by the irregular system of building structure that formerly existed. In addition, I extended the grids towards the interior of the building in different ways. The varied spatial experience so created in the interior was far beyond the thin layer of skin. I also considered and did something on the materials as I

复兴路乙59-1号那个幕墙的不规则网格，是因为原有建筑结构系统的不规则而自然产生的；另外我把这种网格向建筑内部不同的延伸，形成了建筑内部不同的空间体验，而不只是外面的一层薄薄的表皮；同时对材料也有所思考和作为，这也是学习到赫尔佐格与德梅隆关注发掘材料本身潜力的做法，复兴路乙59-1号的幕墙，使用了不同透明度的玻璃，不同的透明度对里面空间有着不同的暗示，同时里面的人，则透过不同通透性的界面，而有不同的景观体验。

黄：建筑是一个视觉化的形体，若您没有去解释这方面的设计过程，其实人家看复兴路乙59-1号改造，因它的外在形象太强烈了，从外面一看，其实就会有倾向于表象或表皮设计的联想。

李：是，但我自己很坦然，因为实际上我不是用那种肤浅的方式做的，我很自信。但唯一觉得遗憾的是，若一个建筑要让人家真正转变那种视觉化的肤浅认识的话，就得让大家去亲身体验，但复兴路这个房子一建成就被业主租出去变成饭馆了，里面的空间改了很多、很差，多半已经不是我们原来的设计，这也是我感到无奈的地方。

黄：如果我们回归到建筑学或者建筑史的视点，如果说尔后人家会去写建筑史的话，很有可能就会把您归在表象或表皮倾向的这一类建筑师当中，就是因为从"鸟巢"到复兴路乙59-1号改造，再到北京地铁四号线出入口，从设计中，可以看到设计表述中的某种一致性与连贯性。

李：事实上"鸟巢"也不是那种"肤浅"的表皮，而是结构化或者空间化的"表皮"……当然我觉得一定会受到影响，因为我是在做"鸟巢"的同时，做这几个项目的，每天都在跟赫尔佐格与德梅隆以及他们的团队打交道，他们是有气场，也会有影响力的，我觉得我不可能完全不受到他们的影响。

黄：就我个人的研究，基本上把表象性归类出四种不同的倾向，分别是"皮层—单面（单层、双层）"的倾向、"皮层—空间"的倾向、"皮层—造型"的倾向与"极少—体量"的倾向。像复兴路乙59-1号改造，就带有一种"皮层—空间"的倾向，利用皮层与建筑内部产生一种空间的联系，形成一种透视延伸的感觉。可是我又觉得复兴路乙59-1号也有一种"皮层—造型"的倾向，它因小见大的重复动作而慢慢形成一种造型，演化成多面向与多角度的整体感，其实"鸟巢"就有点类似这种感觉，它形成一

had learned from HdeM, who focused on mining the potential of materials. The curtain wall on the building was made from glass of different transparencies. The varied transparency would have different implication on the internal space, and the interface of different openness would provide persons in the building with different sightseeing experience.

Huang: A building is a visual body. The Fuxing Road project has such an impressive external image that its appearance tends to remind visitors of a superficial skin design if you don't explain the design process in such aspect.

Li: Yes, that's right. But I feel unperturbed because I didn't do it so superficially. I'm quite confident. But there is only one pity. If we want to change somebody's visually superficial understanding on a building, we have to let him/her physically experience it. However, the Fuxing Road building was rented out by its owner and changed to a restaurant as soon as it was completed. The interior space has been much and badly modified. Most of the design has not been ours any more. I can do nothing about it.

Huang: If we return to the viewpoint of architectonics or architectural history, that is, if somebody writes architectural history in future, it will be quite possible that you are classified as an architect with skin tendency because of the observable consistency in the design representation from "Bird's Nest" to the Fuxing Road project and Accesses of Line 4 of Beijing Subway.

Li: In fact, "Bird's Nest" is not of "superficial" skin, either; instead, it is of structured or spatial "skin"... Certainly, I feel I must have been influenced to some extent because I was doing these projects when at the same time I worked on the "Bird's Nest" project with HdeM and their team every day. They had an atmosphere that must have somewhat affected me.

Huang: In my personal research, I generally classify "Representation" into four tendencies, namely, skin-uniface (single- / double-layer) tendency, skin-space tendency, skin-shape tendency, and rareness-mass tendency. The Fuxing Road project tends to be of the skin-space type because it generates a spatial relationship between the skin and interior to create a feel of perspective. However, I also feel a skin-shape tendency in this building, whose shape is accumulated from repeated minor actions and evolves into a multi-directional and multi-angular integration. In fact, this is the feel similar to that of "Bird's Nest"-a shaping feel. This is why I said you have two routes in design: the Chinese gardens you early focused on and the presentational skin you slightly turned to later. Most recently, as I see, you returned to the early route-gardens-in Jianchuan Mirror Museum & Wenchuan Earthquake Memorial. But you used bricks to treat the skin on the facade to create brick wall surface of different openness. So I feel that you began to combine the two routes in the Jianchuan project. Is that true?

种造型的感觉。所以说，我之前提到您设计上有两条路线，一个是早期关注到中国园林的部分，之后稍稍转向关注到表象、表皮，而近期您在四川安仁建川镜鉴博物馆暨汶川地震纪念馆中，我观察到您又回归到更早的路线——园林，可是在立面上又有用砖来做表皮的处理，塑造出不同通透感的砖砌墙面。所以，我觉得在这个项目中，您开始想把这两条路线做一个结合，是不是这个样子？

李：对，其实复兴路乙59-1号已经有这些结合。复兴路乙59-1号内由西侧防火楼梯改造扩展而成的立体画廊，我把它想象成一个垂直方向的园林，有行走的路径，也有驻停之处，而行动和停留的外表皮透明度不同，比如相当于"亭"的地方完全透明，而在相当于"廊"的地方则是半透明的，就是说表皮跟路径结合起来，跟视觉感官内的景观画面发生了关联，也就是把我关注的园林和"空间表皮"这两个方面结合在一起。建川镜鉴博物馆暨汶川地震纪念馆也是类似，就是说那个砖墙的表皮，它如何来跟里面的空间对应，但是用砖的砌法而不是玻璃的釉面，来做不同的透明度，或者不同的通透度，我们还设计了一种简单便宜的钢板玻璃砖，来代替围合室内空间的那些砖块之间的孔洞；但是博物馆的核心仍然在空间方面，这个空间是一种游园式的空间。

黄：有一种游园象征性的空间。

李：在游园式的空间里，你说的这种象征性，又跟人们对文革的象征性体验联系在一起，所以就形成一种更为复杂的关系和状态。我的设计里面经常有一种让我有些苦恼的特征，就是复杂性。我在一个设计开始的时候总是希望做得简单些，轻松些，但最后的结果总是很复杂，几乎就像失控一样……比如建川博物馆，实际上就有多个方面的思想和表现：第一个方面是园林式的体验，它有狭窄的不起眼的入口，经过庭院和转折，再进入建筑内部，参观展品之后则在水院池边静思感悟并由出口离开；而在内部参观展览的过程中，也有园林式的体验，包括园林空间元素的直接使用——比如复廊，并转化成一种线性的、迂回的展览空间和游览方式。这都是对园林的思考和引用。

第二个方面就像你刚才说到的，建筑的材料和表皮的研究，以及与内部空间的对应性的思考，为什么外面用砖？是因为想形成一种内向性，来突出内部庭院的景观和体验。所以，透明的地方都在里面，外面用砖来封闭，甚

Li: That's right. In fact, they were combined as early as in the Fuxing Road project, where the outdoor emergency stair on the west side was modified and expanded into a stereoscopic gallery. In my imagination, the gallery should be a vertical garden consisting of a walkway and stops. The exterior skin where the space is for movement has a different transparency from that for stops. For example, the skin at places equivalent to "pavilions" should be transparent and the skin at places equivalent to "corridors" should be translucent. That is, the skin is combined with the walkway and the visually sensible landscape pictures; or, the two aspects in my focus, i.e., the garden and the "spatial skin", are integrated with each other. The case is similar to the Jianchuan Museum & Memorial. To fit the brick wall skin with the interior space, bricklaying techniques rather than glazes of glass was used to create different transparency or openness. In addition, we designed the simple and cheap "steel plate + glass bricks", which were used in place of the holes among the bricks that enclosed the interior space. However, the core of the museum is still in its space-a space of garden tour.

Huang: The space symbolizes garden tour.

Li: It is a space of garden tour, while the symbolism you say is related to people's symbolic experience to the Chinese Culture Revolution. Thus, a more complicated relationship and state are formed. Complexity often puzzles me in my design. At the beginning of a design, I always hope to do a simple and easy job. However, the final result is always complex, just like things are out of control. For example, the Jianchuan Museum contains multiple aspects of thoughts and presentations. The first aspect is to experience the museum like visiting a garden: starting from a narrow and unnoticeable entrance, walking through a courtyard and a zigzag path, entering the building, thinking and getting inspired beside a pool in the courtyard after visiting exhibits, and finally leaving through the exit. The garden-type experience is also provided when you visit the interior exhibits. Spatial element of Chinese gardens such as "double-corridor" is directly used and converted to a linear and circuitous exhibition space and tour method. This is the thinking and reference to gardens.

The second aspect, just like you said, is about studying the building material and skin, and thinking about their correspondence to the interior space. Why did we use bricks outside? Because we wanted to create an enclosure to highlight the scenes and experience of the internal courtyard. Therefore, all transparent parts are within the building and enclosed from outside. Indeed, the doors of the exterior shops are also made from opaque metal. Based on the general enclosure created by bricks, we attempted to have some hint on the internal space. So we created the skin of "leaky brick-walls" of different openness.

至外面店铺的门都是金属的、不透明的；而在这种用砖所造成的大的封闭性的基础上，又企图有一些对内部空间的暗示，所以产生了不同通透度的砖砌"花墙"表皮。

第三个方面就是对"文革"主题的表达，镜鉴博物馆展陈的是文革时代的大量镜面展品，对此主题的表达在我看来不可回避。虽然我严格来讲应该不算是经历文革时代的人，但对文革有自己的理解，怎么用一个当代的方式来表现它，又跟特定的展品（镜鉴）发生关联？我们设计了一种"旋转镜门"的装置，利用它的光学特性及其与参观者行为的关联，不断混淆和"误导"人们眼中的虚像景观和现实景观，从而在不知不觉中得到特定主题的体验。

后来加入的汶川地震纪念馆，则完成了地震馆之粗糙"现实感"与镜鉴馆之纯净"幻像感"的内部空间对比，两者既相互独立并置又彼此呼应关联，共同完成特殊历史记忆的营造。这是第四个方面。

这诸多方面交织在一起，造成了这个房子的极大复杂性，虽然它也只有6000多平方米。复杂性，或许也是目前阶段我的设计的一个特征。

黄：我想做一个"实"与"虚"的对比，我观察您在兴涛接待展示中心阶段，还是属于用"实"的阶段，是用实墙去塑造转折与带有点园林的感觉，可是到了建川博物馆与纪念馆的时候，您转到一种"虚"的阶段，比如说用旋转镜门及潜望镜的原理，制造出一系列反射与折射的景象，其实是更抽象化的步换景移手法的再现，所以，您从由一种墙的"实"转到一种镜面的"虚"的运用，这"实"转到"虚"的过程与结果，您的看法是如何？你自己有这个过程吗？

李：当然我觉得这并不是一个有意识的变化，实际上是跟文革镜鉴馆的主题有关系。因为文革在我看来几乎就是一种梦魇般的景象，当时的人们受到某种政治化幻像的吸引和强迫，而不自觉卷入到一种疯狂的群体游戏当中。如果把它变成某种物质化的表达，什么才能造成虚幻的景象？镜子就是一个最好的媒介，而这个馆的展品就是文革镜面，把镜子装置化来表现这种虚幻的氛围，而且真的能把参观转化成游戏，一种始于幻像的游戏活动，让人们亲身体会这种疯狂和迷乱，以此方式提示"以史为鉴"，如果弄得好的话应该是一个挺切题的做法。

The third aspect is expressing the theme of Cultural Revolution. Because the museum would exhibit and display many mirrors from the Cultural Revolution, I didn't think it possible to avoid the expression of such a theme. Although I'm not a person who has truly experienced such an age, I have my own understanding on it. To express it in a contemporary manner and associate it to the specific exhibits (mirrors), we designed a "rotable mirror-door" device. With the optical characteristics of the device and its relationship with visitors' behaviors, people are continuously confused and "misled" by the visualized virtual scene and real scene, and unconsciously experience specific themes.

When Wenchuan Earthquake Memorial was added later on, its interior space provides a rough "feeling of reality" that is in sharp contrast to the pure "vision feeling" of the Mirror Museum. Although the memorial and the museum are independent from each other, they play in concert with one other to create special history memory. This is the forth aspect.

All these aspects mingled with one another and created extreme complexity in this building, which was only a little more than 6,000m^2. Complexity may be one of the characteristics of my present designs.

Huang: I'd like to make a comparison between substantiality and emptiness. As I observed, you were using substantiality in the stage of Xingtao Reception & Exhibition Center, that is, you used substantial walls to create transition and the feel of a garden. However, when it came to the Jianchuan Museum & Memorial, you changed to the stage of emptiness. For example, you used rotable mirror-doors and the periscope principle to make a series of reflected and refracted images. In fact, this was the reproduction of a more abstract technique to change the scene together with the viewer's movement. Therefore, you have converted from the substantial wall to the empty mirror. How do you think about the process and result of such conversion? Have you experienced such a process yourself?

Li: Certainly I don't think this was a deliberate change. In fact, it was related to the theme of the Mirror Museum. For me, the Cultural Revolution was almost a nightmare in which a political illusion absorbed and forced people to unconsciously participate in a crazy and collective game. To materially express the mere illusion, mirrors are the best medium to create illusory scenes. If we build devices using the mirrors in the collection of the museum, which are those from the Cultural Revolution, we can express an illusory atmosphere and truly convert the visit into a game. Beginning from illusion, the game enables people to physically experience the craziness and bewilderment and reminds them to learn from history. This should be a pertinent approach if it is well implemented.

在我的建筑里，不论是对材料的思考，还是对园林的思考，或是中国性空间与语言的思考，我都不希望它是一个先入为主的观念，就是跟建筑的功能、主题与内在逻辑、需求完全不关联的那种，我希望能够把所有的它们都很自然地联接起来。就是说，是一种内在的需求，让你产生这样一种材料的使用，这样一种空间的存在，这样一种景象的营造。我觉得最好它是一种建筑自己的愿望，对我来讲，这才是心中理想的、美好的建筑。

黄：您从早期到现在，这样一系列项目的思考与操作，我想问您有没有自己设计的中心思想或者信仰？

李：可能有十几年了，我被大家叫做"青年建筑师"，它的一个不好的地方，是让我总会对自己的某种放任。回顾我之前的这些工作，我觉得有一种"想到哪做到哪"那么一种心态，我没有特别有意识地去建立自己的比如说语言体系或者什么，像你说的，主要建筑观之类的。

黄：中心思想之类的。

李：我肯定也有，但我没有有意识地把它强化或者凸显出来，有时候迫于媒体要求，挤出几句，也是笼而统之。我认为主动和有意识还是很重要的，现在开始慢慢觉得需要有这样一种意识。刚才为什么说我做设计有点想到就做到哪，对自己有些放任，就是我觉得做这个项目的时候，可能对这个方面我有兴趣，然后正好是有碰触点的，有结合的可能，那我可能就这么做一下；而做那个项目的时候，我又会做那个方面，我自己做得也挺愉快。但是可能也会有一种苦恼，就是说因为你没有有意识地去思考某种语言系统，或者某种目标性的课题，可能就不能够集中能量和精力去研究、去强化并得到某种代表性的成果吧。

那么我觉得可能在未来的时间里，也许我需要有意识地开始。因为现在年过四十，即便作为老人职业的建筑师，也已经不再年轻，也有一点压力吧，所以要有意识地再成熟一些，需要有一些主动点的努力。

但是我也觉得，这个不是说你想这样你就能这样，而是需要有一个自然而然的过程，需要你的经验，你的思考，你的某种境界，你能否到得了那个程度，可能才会自然而然地出现；如果不能出现，说明还没到火候，或者干脆我就没有这样的能力。

I don't hope any of my thoughts about materials, gardens, or about any space and language with Chinese characteristics is so preconceived that it is irrelevant to any function, theme, internal logic, or need of the buildings; rather, I expect that all these elements can be naturally connected to each other. That is to say, an inherent need lets you use a material in such a way, create such a space, or build such a scene. I think the best thing is out of the own desire of the building. Something like this is the ideal and perfect architecture in my mind.

Huang: From your early stage to the present, you have thought about and operated such a series of projects. I wonder if you have your central idea or belief in your design.

Li: Perhaps people have called me "a young architect" for more than a decade. This has a side effect that I always indulge myself. Looking back on my previous work, I feel that I haven't consciously established, for example, my language system or, as you say, a primary outlook on architecture.

Huang: Central idea or alike.

Li: I must have it but haven't consciously enhanced or highlighted it. Sometimes I may squeeze out a few words in response to media requests and contain nothing specific. In fact, I think it important to be proactive and have begun to realize that I need such consciousness. As I said, I indulge myself in my design because, in one project, I may do something that interests me and is possible to be incorporated; but in another project, I may do something else. Although I enjoy such a practice, there may be a worry. Because you don't consciously think about a certain language system or about subjects with a certain objective, you may feel it impossible to concentrate your energy and effort to study and enhance something and get a representative result.

So I feel I may need to consciously start something in future. There is some pressure because I, more than 40 years old, am not young any more even as an architect who is "the older, the better". This is why I should proactively and consciously make myself more mature.

But this doesn't necessarily mean that you can get it as soon as you want it before going through a natural process. It needs your experience, thinking, and a certain degree of perfection. It may naturally appear if you reach that degree; and if it doesn't, it's not the time or I do not have the ability at all.

作品年表　　　　　　　　　　　　　　Chronology of Works

★──文中收录作品　　　　　　　★── Selected Works in Text
●──实现作品　　　　　　　　　●── Built Works
◐──实现中作品　　　　　　　　◐── Works in Construction
○──未实现作品　　　　　　　　○── Unbuilt Works

项目名称：兴涛社区 / ●
Project: Xingtao Residencial Quarter
设计团队：李兴钢 高为 林瞳 叶蕾 马先 /
Team: Li Xinggang, Gao Wei, Lin Tong, Ye Lei, Ma Xian
结构设计：李淑捧 张晔 王力波 /
Structure: Li Shupeng, Zhang Ye, Wang Libo
建设地点 / Location：北京 / Beijing
设计时间 / Design Period：1995.07~2000.04
施工时间 / Construction Period：1996.10~2002.10
用地面积 / Site Area：210900m²
建筑面积 / Building Area：230969m²
获奖情况：北京市优秀工程设计一等奖（2001）等 /
Award: 1st Prize of Beijing Outstanding Architecture Awards (2001), etc.

项目名称：新华大厦 / ●
Project: Xinhua Hotel
设计团队：李兴钢 谭泽阳 于家峰 苏航 马先 张晖 /
Team: Li Xinggang, Tan Zeyang, Yu Jiafeng, Su Hang, Ma Xian, Zhang Hui
结构设计 / Structure：张晔 / Zhang Ye
建设地点 / Location：河北唐山 / Tangshan, Hebei
设计时间 / Design Period：1998.03~1998.10
施工时间 / Construction Period：1998.10~2001.11
用地面积 / Site Area：5400m²
建筑面积 / Building Area：18852m²

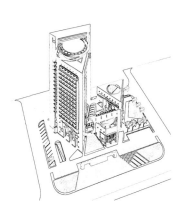

项目名称：泰达小学 / ●
Project: Teda Primary School
设计团队：李兴钢 胡水菁 张晔 宋淑辉 王宇 /
Team: Li Xinggang, Hu Shuijing, Zhang Ye, Song Shuhui, Wang Yu
结构设计 / Structure：刘连荣 / Liu Lianrong
建设地点 / Location：天津 / Tianjin
设计时间 / Design Period：1998.09~2000.06
施工时间 / Construction Period：2000.06~2001.09
用地面积 / Site Area：17141m²
建筑面积 / Building Area：15478m²
获奖情况：中国建筑学会建筑创作大奖提名奖（2009），中国建筑学会建筑创作奖佳作奖（2006）等 /
Award: Finalist of Architectural Society of China Great Design Awards (2009), Outstanding Prize of Architectural Society of China Design Awards (2006), etc.

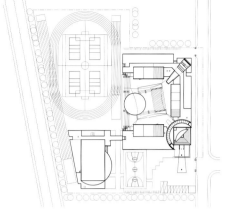 

项目名称：兴涛会馆 / ○
Project: Xingtao Club
设计团队：李兴钢 朱荷蒂 曾宁燕 /
Team: Li Xinggang, Zhu Hedi, Zeng Ningyan
结构设计 / Structure：梁伟 / Liang Wei
建设地点 / Location：北京 / Beijing
设计时间 / Design Period：1999.07~2000.08
用地面积 / Site Area：9214m²
建筑面积 / Building Area：10629m²
获奖情况：首都十佳建筑设计方案奖（2000）等 /
Award: Capital Top-ten Architecture in Beijing (2000), etc.

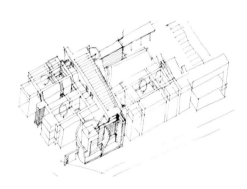 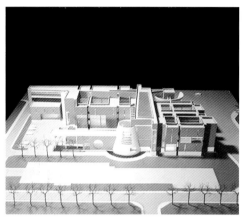

项目名称：西环广场暨西直门交通枢纽（合作）/ ●
Project: Xihuan Plaza & Xizhimen Transportation Exchange Hub (cooperated with AREP, France)
设计团队：李兴钢 魏篙川 苗茁 陈帅飞 杜地阳（法）铁凯歌（法）等 /
Team: Li Xinggang, Wei Gaochuan, Miao Zhuo, Chen Shuaifei, Jean Marie Duthilleul, Etienne Tricand, etc.
结构设计 / Structure：吴平 彭永宏 / Wu Ping, Peng Yonghong
建设地点 / Location：北京 / Beijing
设计时间 / Design Period：2000.05~2007.04
施工时间 / Construction Period：2003.03~2008.03
用地面积 / Site Area：59900m²
建筑面积 / Building Area：252790m²
获奖情况：首都十佳建筑设计方案奖（2001）等 /
Award: Capital Top-ten Architecture in Beijing (2001), etc.

项目名称：中关村生命科学园会展中心竞赛 / ○
Project: Competition for Conference & Exhibition Center of Zhongguancun Biological Science Park
设计团队：李兴钢 陈奕鹏 丁峰 /
Team: Li Xinggang, Chen Yipeng, Ding Feng
建设地点 / Location：北京 / Beijing
设计时间 / Design Period：2001.03
用地面积 / Site Area：62379m²
建筑面积 / Building Area：58653m²

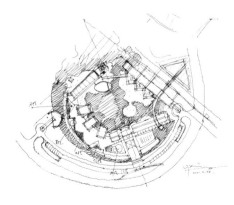 

项目名称：兴涛接待展示中心 / ●★
Project: Xingtao Reception & Exhibition Center
设计团队：李兴钢 李靖 谭泽阳 /
Team: Li Xinggang, Li Jing, Tan Zeyang
结构设计 / Structure：王力波 / Wang Libo
建设地点 / Location：北京 / Beijing
设计时间 / Design Period：2001.04~2001.06
施工时间 / Construction Period：2001.06~2001.09
用地面积 / Site Area：2510m²
建筑面积 / Building Area：883m²
获奖情况：中国建筑艺术奖（公共建筑类）（2003）、英国世界建筑奖提名奖（2002）/
Award: China Architecture Art Awards (Public Building)(2003), Finalist of World Architecture Awards of UK (2002)

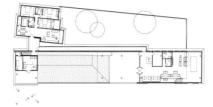

项目名称：大兴文化中心 / ●
Project: Daxing Culture Center
设计团队：李兴钢 陈晓唐 张晔 陈泽勇 /
Team: Li Xinggang, Chen Xiaotang, Zhang Ye, Chen Zeyong
结构设计 / Structure：孔雅莎 / Kong Yasha
建设地点 / Location：北京 / Beijing
设计时间 / Design Period：2001.07~2004.09
施工时间 / Construction Period：2003.06~2006.06
用地面积 / Site Area：17390m²
建筑面积 / Building Area：28620m²
获奖情况：中国建筑学会建筑创作奖佳作奖（2008）、全国工程勘察设计行业优秀工程二等奖（2008）等 /
Award: Outstanding Prize of Architectural Society of China Design Awards (2008), 2nd Prize of China Architecture & Prospecting Trades Outstanding Architecture Awards (2008), etc.

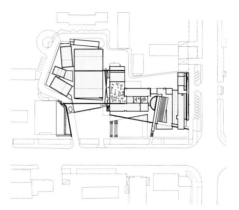

项目名称：东莞理工学院科研测试中心及后勤服务用房 / ◎
Project: Science Research Center & Service Facilities of Dongguan Technology Institute
设计团队：李兴钢 付邦保 李大丹 肖晓丽 徐群宁 /
Team: Li Xinggang, Fu Bangbao, Li Dadan, Xiao Xiaoli, Xu Qunning
结构设计 / Structure：陈刘刚 / Chen Liugang
建设地点 / Location：广东东莞 / Dongguan, Guangdong
设计时间 / Design Period：2002.09~2003.01
施工时间 / Construction Period：2003.01至今
用地面积 / Site Area：34845m²
建筑面积 / Building Area：26612m²
获奖情况：上海国际青年建筑师设计作品展一等奖（2004）/
Award: 1st Prize of Shanghai International Youth Architect Works Exhibition (2004)

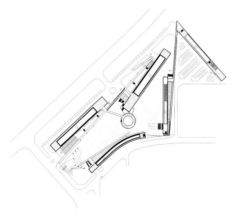
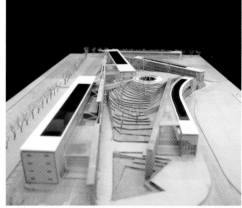

项目名称：松山湖文化营多功能活动中心 / ○
Project: Multi-Center of Songshanhu Culture Camp
设计团队：李兴钢 付邦保 张音玄 /
Team: Li Xinggang, Fu Bangbao, Zhang Yinxuan
建设地点 / Location：广东东莞 / Dongguan, Guangdong
设计时间 / Design Period：2003.08~2003.11
建筑面积 / Building Area：17130m²

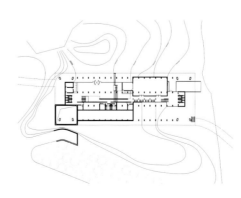

项目名称：湖北省艺术馆竞赛 / ○
Project: Competition for Hubei Art Gallery
设计团队：李兴钢 付邦保 张音玄 /
Team: Li Xinggang, Fu Bangbao, Zhang Yinxuan
建设地点 / Location：湖北武汉 / Wuhan, Hubei
设计时间 / Design Period：2003.11
用地面积 / Site Area：20417m²
建筑面积 / Building Area：22654m²
获奖情况：上海国际青年建筑师设计作品展二等奖（2004）/
Award: 2nd Prize of Shanghai International Youth Architect Works Exhibition (2004)

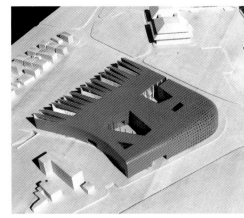

项目名称：建川文革镜鉴博物馆暨汶川地震纪念馆 / ●★
Project: Jianchuan Mirror Museum & Wenchuan Earthquake Memorial
设计团队：李兴钢 张音玄 付邦保 谭泽阳 刘爱华 闫昱 等 /
Team: Li Xinggang, Zhang Yinxuan, Fu Bangbao, Tan Zeyang, Liu Aihua, Yan Yu, etc.
结构设计 / Structure：王力波 余蕾 / Wang Libo, Yu Lei
建设地点 / Location：四川安仁 / Anren, Sichuan
设计时间 / Design Period：2004.02~2009.12
施工时间 / Construction Period：2004.08~2010.09
用地面积 / Site Area：3847m²
建筑面积 / Building Area：6098m²
获奖情况：THE CHICAGO ATHENUM国际建筑奖（2010）、欧洲维纳博格奖金砖奖提名奖（2010）等 /
Award: THE CHICAGO ATHENUM International Architecture Awards (2010), Nomination of Weinerberger Golden Brick Awards (2010), etc.

项目名称：复兴路乙59-1号改造 / ●★
Project: Reconstruction of No.B-59-1, Fuxing Road
设计团队：李兴钢 张音玄 付邦保 谭泽阳 等 /
Team: Li Xinggang, Zhang Yinxuan, Fu Bangbao, Tan Zeyang, etc.
结构设计 / Structure：蒋航军 / Jiang Hangjun
建设地点 / Location：北京 / Beijing
设计时间 / Design Period：2004.10~2005.08
施工时间 / Construction Period：2005.07~2007.05
用地面积 / Site Area：1280m²
建筑面积 / Building Area：5402m²
获奖情况：全国优秀工程设计银奖（2010）、THE CHICAGO ATHENUM国际建筑奖（2009）等 /
Award: Silver Prize of China National Outstanding Architecture Awards (2010), THE CHICAGO ATHENUM International Architecture Awards (2009), etc.

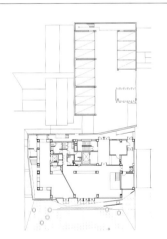

项目名称：中华英才半月刊社综合业务楼 / ◎
Project: Office Building of Top China Microdevices
设计团队：李兴钢 张音玄 付邦保 郭佳 /
Team: Li Xinggang, Zhang Yinxuan, Fu Bangbao, Guo Jia
建设地点 / Location：北京 / Beijing
设计时间 / Design Period：2005.12至今
用地面积 / Site Area：4528m²
建筑面积 / Building Area：13332m²

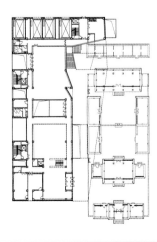

项目名称：Hiland·名座 / ●★
Project: Hiland·Mingzuo
设计团队：李兴钢 谭泽阳 李宁 钟鹏 肖育智 等 /
Team: Li Xinggang, Tan Zeyang, Li Ning, Zhong Peng, Xiao Yuzhi, etc.
结构设计：尤天直 唐杰 高文军 /
Structure: You Tianzhi, Tang Jie, Gao Wenjun
建设地点 / Location：山东威海 / Weihai, Shandong
设计时间 / Design Period：2006.01~2007.09
施工时间 / Construction Period：2007.10~2010.05
用地面积 / Site Area：3618m²
建筑面积 / Building Area：26190m²
获奖情况：中国建筑学会建筑创作奖佳作奖（2011）等 /
Award: Outstanding Prize of Architectural Society of China Design Awards (2011), etc.

项目名称：乳山文博中心 / ○
Project: Rushan Culture Center & Museum
设计团队：李兴钢 张哲 钟鹏 弓蒙 /
Team: Li Xinggang, Zhang Zhe, Zhong Peng, Gong Meng
建设地点 / Location：山东乳山 / Rushan, Shandong
设计时间 / Design Period：2006.07~2007.06
用地面积 / Site Area：560000m²
建筑面积 / Building Area：24000m²

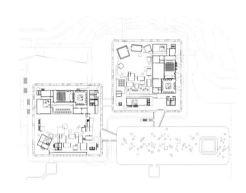
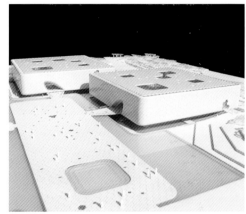

项目名称：岢园 / ○
Project: Ke Garden
设计团队：李兴钢 付邦保 张哲 郭佳 /
Team: Li Xinggang, Fu Bangbao, Zhang Zhe, Guo Jia
建设地点 / Location：广西南宁 / Nanning, Guangxi
设计时间 / Design Period：2006.12~2008.01
用地面积 / Site Area：3389m²
建筑面积 / Building Area：1246m²

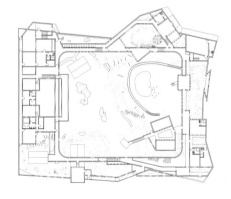
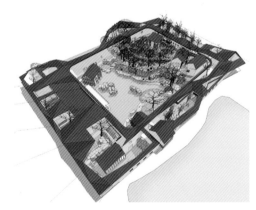

项目名称：乐高一号 / ●
Project: Lego I
设计团队：李兴钢 张音玄 付邦保 张哲 /
Team: Li Xinggang, Zhang Yinxuan, Fu Bangbao, Zhang Zhe
建设地点 / Location：北京 / Beijing
设计时间 / Design Period：2007.04
施工时间 / Construction Period：2007.07
作品尺寸 / Size：400mm × 500mm × 1200mm
参展情况：北京大声艺术展（2007）、深圳城市建筑双年展（2007）/
Exhibition: "Get It Louder", Beijing (2007), 2nd Shenzhen Biennale of Urbanism & Architecture (2007)

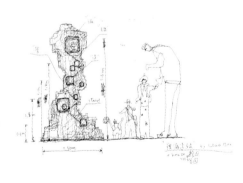

项目名称：唐山地震遗址纪念公园竞赛 / ○
Project: Competition for Tangshan Earthquake Site Memorial Park
设计团队：李兴钢 郭佳 张哲 李宁 弓蒙 /
Team: Li Xinggang, Guo Jia, Zhang Zhe, Li Ning, Gong Meng
建设地点 / Location：河北唐山 / Tangshan, Hebei
设计时间 / Design Period：2007.05~2007.07
用地面积 / Site Area：400000m²

项目名称：中国海关博物馆（合作）/ ○
Project: China Custom Museum (cooperated with Approach Architecture & Jiuyuan Samsung Architects)
设计团队：李兴钢 梁井宇（合作）付邦保 肖育智 彭小虎 江曼 张安星 等 /
Team: Li Xinggang, Liang Jingyu (Cooperated), Fu Bangbao, Xiao Yuzhi, Peng Xiaohu, Jiang Man, Zhang Anxing, etc.
结构设计：李欣 /
Structure: Li Xin
建设地点 / Location：北京 / Beijing
设计时间 / Design Period：2007.06~2009.10
施工时间 / Construction Period：2009.11至今
用地面积 / Site Area：21023m²
建筑面积 / Building Area：33000m²

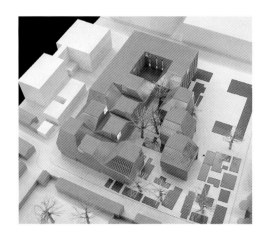

项目名称：深圳湾体育中心竞赛 / ○
Project: Competition for Shenzhen Bay Sports Center
设计团队：李兴钢 付邦保 钟鹏 李宁 邢迪 王子耕 谭泽阳 /
Team: Li Xinggang, Fu Bangbao, Zhong Peng, Li Ning, Xing Di, Wang Zigeng, Tan Zeyang
建设地点 / Location：广东深圳 / Shenzhen, Guangdong
设计时间 / Design Period：2007.09~2007.11
用地面积 / Site Area：307740m²
建筑面积 / Building Area：268621m²

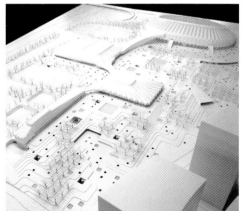

项目名称：藕园 / ○
Project: Couple Garden
设计团队：李兴钢 付邦保 邢迪 /
Team: Li Xinggang, Fu Bangbao, Xing Di
建设地点 / Location：广东番禺 / Panyu, Guangdong
设计时间 / Design Period：2007.10~2008.01
用地面积 / Site Area：985m²
建筑面积 / Building Area：1270m²

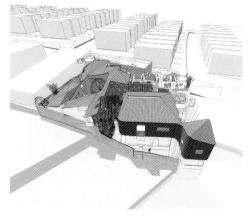

项目名称：乐高二号 / ●
Project: Lego II
设计团队：李兴钢 付邦保 郭佳 李宁 邢迪 张玉婷 /
Team: Li Xinggang, Fu Bangbao, Guo Jia, Li Ning, Xing Di, Zhang Yuting
建设地点 / Location：北京 / Beijing
设计时间 / Design Period：2008.03
施工时间 / Construction Period：2008.04
作品尺寸 / Size：1800×1000×1200 mm
参展情况：德累斯顿"从幻象到现实：活的中国园林"展（2008）/
Exhibition: "Illusion Into Reality: Chinese Gardens for Living", Dresden (2008)

项目名称：纸砖房 / ●★
Project: Paper-brick House
设计团队：李兴钢 付邦保 李宁 孙鹏 等 /
Team: Li Xinggang, Fu Bangbao, Li Ning, Sun Peng, etc.
建设地点 / Location：意大利威尼斯 / Venice, Italy
设计时间 / Design Period：2008.05
施工时间 / Construction Period：2008.09
作品尺寸 / Size：15m×2m×4m
参展情况：第11届威尼斯国际建筑双年展中国馆（2008）/
Exhibition: Chinese Pavilion of 11th Venice Biennial of Architecture (2008)
获奖情况：THE CHICAGO ATHENUM国际建筑奖（2009）/
Award: THE CHICAGO ATHENUM International Architecture Awards (2009)

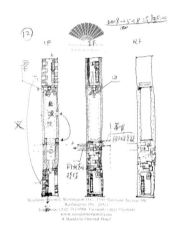

项目名称：李兴钢工作室室内设计 / ●★
Project: Interior Design of Atelier Li Xinggang
设计团队：李兴钢 郭佳 张玉婷 邱涧冰 谭泽阳 /
Team: Li Xinggang, Guo Jia, Zhang Yuting, Qiu Jianbing, Tan Zeyang
建设地点 / Location：北京 / Beijing
设计时间 / Design Period：2008.07~2008.09
施工时间 / Construction Period：2008.09~2009.02
建筑面积 / Building Area：485m²

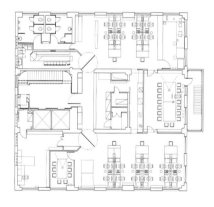
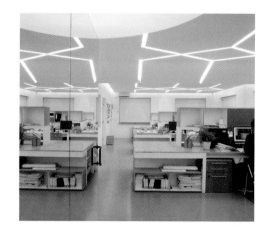

项目名称：北京地铁4号线及大兴线地面出入口及附属设施 / ●★
Project: Accesses and subsidiaries of Line 4 & Daxing Line of Beijing Subway
设计团队：李兴钢 张音玄 邱涧冰 肖育智 李宁 邢迪 张玉婷 /
Team: Li Xinggang, Zhang Yinxuan, Qiu Jianbing, Xiao Yuzhi, Li Ning, Xing Di, Zhang Yuting
结构设计 / Structure：高银鹰 / Gao Yinying
建设地点 / Location：北京 / Beijing
设计时间 / Design Period：2008.08~2009.03|2010.02~2010.04
施工时间 / Construction Period：2009.03~2009.10|2010.07~2010.12
建筑面积 / Building Area：26190m²|8371m²
（4号线|大兴线 / Line 4|Daxing Line）
获奖情况：北京市优秀工程设计一等奖（2011）/
Award: 1st Prize of Beijing Outstanding Architecture Awards (2011)

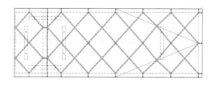
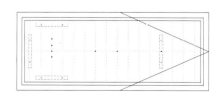

项目名称：北京地铁昌平线西二旗站 / ●★
Project: Xi'erqi Station of Changping Line of Beijing Subway
设计团队：李兴钢 张音玄 张哲 张玉婷 李慧琴 赵国璆 /
Team: Li Xinggang, Zhang Yinxuan, Zhang Zhe, Zhang Yuting, Li Huiqin, Zhao Guoqiu
结构设计 / Structure：王力波 / Wang Libo
建设地点 / Location：北京 / Beijing
设计时间 / Design Period：2008.11~2010.04
施工时间 / Construction Period：2009.10~2010.12
建筑面积 / Building Area：16670m²
获奖情况：北京国际设计周年度建筑设计奖（2011）、国际膜结构协会杰出贡献奖（2011）等 /
Award: Beijing Design Week Annual Award for Architecture Design (2011), IFAI Outstanding Achievement Award for Tensile Structures (2011), etc.

项目名称：商丘博物馆 / ○★
Project: Shangqiu Museum
设计团队：李兴钢 谭泽阳 付邦保 郭佳 张哲 李喆 等 /
Team: Li Xinggang, Tan Zeyang, Fu Bangbao, Guo Jia, Zhang Zhe, Li Zhe, etc.
结构设计：王力波 张晔 杨威 /
Structure: Wang Libo, Zhang Ye, Yang Wei
建设地点 / Location：河南商丘 / Shangqiu, Henan
设计时间 / Design Period：2008.12~2010.09
施工时间 / Construction Period：2011.03至今
用地面积 / Site Area：73613m²
建筑面积 / Building Area：29672m²

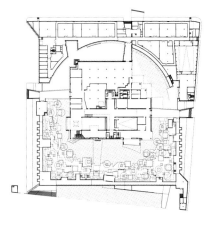 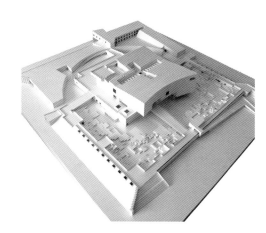

项目名称："第三空间" / ○★
Project: "The Third Space"
设计团队：李兴钢 付邦保 孙鹏 赵小雨 谭泽阳 张一婷 等 /
Team: Li Xinggang, Fu Bangbao, Sun Peng, Zhao Xiaoyu, Tan Zeyang, Zhang Yiting, etc.
结构设计 / Structure：张付奎 孔文华 / Zhang Fukui, Kong Wenhua
建设地点 / Location：河北唐山 / Tangshan, Hebei
设计时间 / Design Period：2009.02~2010.09
施工时间 / Construction Period：2010.06至今
用地面积 / Site Area：12852m²
建筑面积 / Building Area：88011m²

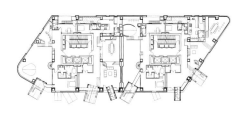 

项目名称：北京地铁6号线隆福寺站周边商业地块概念设计 / ○
Project: Concept Design of Commercial Block of Longfusi Station of Line 6 of Beijing Subway
设计团队：李兴钢 邱涧冰 张玉婷 张音玄 肖育智 张哲 李喆 钟曼琳 /
Team: Li Xinggang, Qiu Jianbing, Zhang Yuting, Zhang Yinxuan, Xiao Yuzhi, Zhang Zhe, Li Zhe, Zhong Manlin
建设地点 / Location：北京 / Beijing
设计时间 / Design Period：2009.03至今
用地面积 / Site Area：12760m²
建筑面积 / Building Area：49895m²

项目名称：元上都遗址博物馆 / ◎★
Project: Museum for Site of XANADU
设计团队：李兴钢 谭泽阳 付邦保 赵小雨 /
Team: Li Xinggang, Tan Zeyang, Fu Bangbao, Zhao Xiaoyu
结构设计 / Structure：王力波 高银鹰 / Wang Libo, Gao Yinying
建设地点：内蒙古正蓝旗 /
Location: Zhenglanqi, Inner Mongolia
设计时间 / Design Period：2009.05~2010.07
施工时间 / Construction Period：2010.07至今
用地面积 / Site Area：6252m²
建筑面积 / Building Area：5701m²

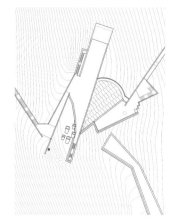

项目名称：海南国际会展中心 / ●★
Project: Hainan International Conference & Exhibition Center
设计团队：李兴钢 谭泽阳 付邦保 张玉婷 /
Team: Li Xinggang, Tan Zeyang, Fu Bangbao, Zhang Yuting
结构设计：任庆英 王载 王文宇 谷昊 /
Structure: Ren Qingying, Wang Zai, Wang Wenyu, Gu Hao
三维协作 / 3D Support：徐东昕 李铮 / Xu Dongxin, Li Zheng
建设地点 / Location：海南海口 / Haikou, Hainan
设计时间 / Design Period：2009.07~2010.03
施工时间 / Construction Period：2009.11~2011.07
用地面积 / Site Area：319873m²
建筑面积 / Building Area：132788m²

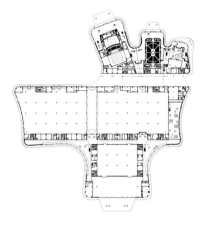

项目名称：渤龙湖总部基地二区 / ◎
Project: Section 2 of Bolonghu Headquarter Base
设计团队：李兴钢 邱涧冰 李宁 薛从清 肖育智 闫昱 唐勇 等 /
Team: Li Xinggang, Qiu Jianbing, Li Ning, Xue Congqing, Xiao Yuzhi, Yan Yu, Tang Yong, etc.
结构设计 / Structure：张付奎 王载 / Zhang Fukui, Wang Zai
建设地点 / Location：天津 / Tianjin
设计时间 / Design Period：2009.07至今
施工时间 / Construction Period：2009.12至今
用地面积 / Site Area：192726m²
建筑面积 / Building Area：297111m²

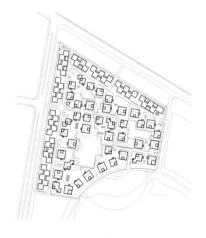

项目名称：绩溪博物馆 / ◎★
Project: Jixi Museum
设计团队：李兴钢 张音玄 张哲 张一婷 易灵洁 /
Team: Li Xinggang, Zhang Yinxuan, Zhang Zhe, Zhang Yiting, Yi Lingjie
结构设计 / Structure：王力波 杨威 / Wang Libo, Yang Wei
建设地点 / Location：安徽绩溪 / Jixi, Anhui
设计时间 / Design Period：2009.11~2010.12
施工时间 / Construction Period：2010.12至今
用地面积 / Site Area：9500m²
建筑面积 / Building Area：10003m²

项目名称：鄂尔多斯20+10项目D4、P19地块设计 / ○
Project: Ordos 20+10 Project D4 & P19
设计团队：李兴钢 谭泽阳 邢迪 唐勇 /
Team: Li Xinggang, Tan Zeyang, Xing Di, Tang Yong
结构设计 / Structure：王多民 赵岩 / Wang Duoming, Zhao Yan
建设地点 / Location：内蒙古鄂尔多斯 / Erdos, Inner Mongolia
设计时间 / Design Period：2010.02~2011.06
用地面积 / Site Area：10440m²
建筑面积 / Building Area：33053m²

 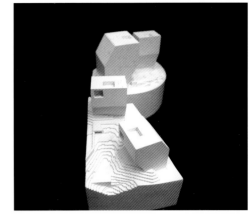

项目名称：MAX LAB IV设计竞赛（合作） / ○
Project: MAX LAB IV Competition (cooperated with FOJAB, Sweden)
设计团队：李兴钢 张玉婷 易灵洁 乌尔夫（瑞典） 芮歇尔（瑞典） /
Team: Li Xinggang, Zhang Yuting, Yi Lingjie, Ulf Kadefors, Rachelle Astrand
建设地点 / Location：瑞典兰德 / Lund, Sweden
设计时间 / Design Period：2010.07~2010.10
用地面积 / Site Area：197687m²
建筑面积 / Building Area：37860m²

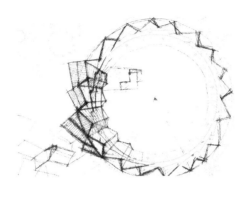 

项目名称：元上都遗址工作站 / ●★
Project: Entrance for Site of XANADU
设计团队：李兴钢 邱涧冰 易灵洁 孙鹏 张玉婷 赵小雨 /
Team: Li Xinggang, Qiu Jianbing, Yi Lingjie, Sun Peng, Zhang Yuting, Zhao Xiaoyu
结构设计 / Structure：高银鹰 / Gao Yinying
建设地点：内蒙古正蓝旗
Location: Zhenglanqi, Inner Mongolia
设计时间 / Design Period：2010.08~2011.05
施工时间 / Construction Period：2011.05~2011.08
用地面积 / Site Area：16653m²
建筑面积 / Building Area：410m²
参展情况：伦敦"从北京到伦敦—当代中国建筑"展（2012）/
Exhibition:"From Beijing to London: 16 Contemporary Chinese Architects", London (2012)

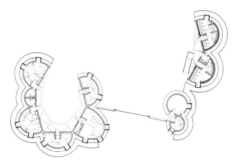 

项目名称：西柏坡华润希望小镇 / ○★
Project: Xibaipo China Resources Hope Town
设计团队：李兴钢 谭泽阳 邱涧冰 梁旭 张一婷 马津 赵小雨 等 /
Team: Li Xinggang, Tan Zeyang, Qiu Jianbing, Liang Xu, Zhang Yiting, Ma Jin, Zhao Xiaoyu, etc.
结构设计：毕磊 何羽 何喜明 /
Structure: Bi Lei, He Yu, He Ximing
建设地点 / Location：河北西柏坡 / Xibaipo, Hebei
设计时间 / Design Period：2010.09至今
施工时间 / Construction Period：2011.03至今
用地面积 / Site Area：153800m²
建筑面积 / Building Area：53100m²

项目名称：北京CBD核心区城市设计 / ○
Project: Urban Design of Core Area of Beijing CBD
设计团队：李兴钢 张玉婷 李喆 易灵洁 /
Team: Li Xinggang, Zhang Yuting, Li Zhe, Yi Lingjie
建设地点 / Location：北京 / Beijing
设计时间 / Design Period：2010.09~2010.11
用地面积 / Site Area：246720m²
建筑面积 / Building Area：2000000m²

项目名称：通州新城运河核心区VII09-14用地概念设计 / ○
Project: Concept Design of Canal Core Area VII09-14, Tongzhou New Town
设计团队：李兴钢 张玉婷 李喆 /
Team: Li Xinggang, Zhang Yuting, Li Zhe
建设地点 / Location：北京 / Beijing
设计时间 / Design Period：2010.10
用地面积 / Site Area：83453m²
建筑面积 / Building Area：700000m²

项目名称：天津大学新校区室内体育活动中心和游泳馆 / ◎
Project: Gymnasium & Natatorium of New Campus of Tianjin University
设计团队：李兴钢 张音玄 闫昱 易灵洁 /
Team: Li Xinggang, Zhang Yinxuan, Yan Yu, Yi Lingjie
结构设计 / Structure：任庆英 / Ren Qingying
建设地点 / Location：天津 / Tianjin
设计时间 / Design Period：2011.02至今
用地面积 / Site Area：75000m²
建筑面积 / Building Area：18362m²

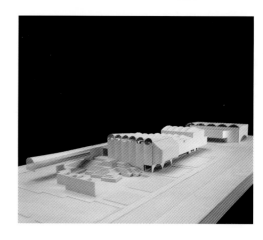

项目名称：中国建筑设计研究院BIM绿色科研办公示范楼竞赛 / ○
Project: Competition for BIM Ecology Office Building of China Architecture Design & Research Group
设计团队：李兴钢 张音玄 张哲 闫昱 唐勇 李喆 /
Team: Li Xinggang, Zhang Yinxuan, Zhang Zhe, Yan Yu, Tang Yong, Li Zhe
建设地点 / Location：北京 / Beijing
设计时间 / Design Period：2011.04~2011.05
用地面积 / Site Area：4114m²
建筑面积 / Building Area：33108m²

■ 项目介绍 Project Description
■ 李兴钢 Li Xinggang

■ 图纸整理 Drawing Reorganization
■ 钟曼琳 Zhong Manlin

■ 版面设计 Layout Design
■ 易灵洁 Yi Lingjie

■ 英文翻译 English Translation
■ 周 萱 Zhou Xuan

■ 文字校对 Text Collation
■ 李兴钢 Li Xinggang
　易灵洁 Yi Lingjie

李兴钢工作室成员 / Atelier Li Xinggang Team

李兴钢　谭泽阳　张音玄　李宁　邢迪　邱涧冰　张哲　张玉婷　孙鹏　赵小雨　李喆　闫昱　唐勇　梁旭　易灵洁　马津（研究生）　钟曼琳（研究生）/
Li Xinggang, Tan Zeyang, Zhang Yinxuan, Li Ning, Xing Di, Qiu Jianbing, Zhang Zhe, Zhang Yuting, Sun Peng, Zhao Xiaoyu, Li Zhe, Yan Yu, Tang Yong, Liang Xu, Yi Lingjie, Ma Jin (postgraduate), Zhong Manlin (postgraduate)

曾经成员 / Former Collaborators

付邦保　李力　刘爱华　钟鹏　董煊　肖育智　郭佳　王子耕　薛从清　戴泽钧（研究生）　弓蒙（研究生）　朱磊（研究生）　张一婷（研究生）/
Fu Bangbao, Li Li, Liu Aihua, Zhong Peng, Dong Xuan, Xiao Yuzhi, Guo Jia, Wang Zigeng, Xue Congqing, Dai Zejun (postgraduate), Gong Meng (postgraduate), Zhu Lei (postgraduate), Zhang Yiting (postgraduate)

李兴钢简介

李兴钢,建筑师,工学博士。1969年出生,中国建筑设计研究院(集团)副总建筑师、李兴钢工作室主持人。1991年毕业于天津大学建筑系,1998年入选法国总统项目"50位中国建筑师在法国"在法进修,2012年获得天津大学建筑设计及其理论专业博士学位。曾获得中国青年科技奖(2007)、国务院政府特殊津贴专家(2006)、中国建筑学会青年建筑师奖(2005)、亚洲建筑推动奖(2004)等荣誉;获得THE CHICAGO ATHENUM国际建筑奖(2010/2009)、全国优秀工程设计金/银奖(2009/2010/2000)、全球华人青年建筑师奖(2007)、中国建筑艺术奖(2003)、英国世界建筑奖提名奖(2002)等设计奖项;受邀参加伦敦"从北京到伦敦——当代中国建筑"展(2012)、罗马"向东方——中国建筑景观2011"展(2011)、巴塞尔/列支敦士登"东风-中国新建筑2000-2010"展(2010)、卡尔斯鲁厄/布拉格"后实验时代的中国地域建筑"展(2010)、布鲁塞尔"心造——中国当代建筑的前沿"展(2009)、第11届威尼斯国际建筑双年展(2008)、德累斯顿"从幻象到现实:活的中国园林"展(2008)、北京大声艺术展(2007)、"发生"——北京左右艺术区艺术展(2007)、深圳城市建筑双年展(2007/2005)、"状态"——中国当代青年建筑师作品八人展(2005)等展览。

李兴钢工作室倡导研究性设计与设计性研究的并重和融合,强调建筑对文脉、自然、场地、材料、使用者的反应和对建筑、室内、景观的全过程设计与控制,创作体现当代性并具有文化厚度和艺术感染力的建筑作品,并开展"园林—聚落"和"大匠阅读"等系列专项研究。

Profile

Li Xinggang, architect, Doctor of Engineering, was born in 1969 and is currently the vice chief architect of China Architecture Design & Research Group and the director of Atelier Li Xinggang. In 1991, he graduated from the Department of Architecture at Tianjin University, China. In 1998, he won a scholarship of French President's Program "50 Chinese Architects in France". In 2012, he was awarded a Doctorate in Architectural Design and Theory. He has won various honors and awards of architecture such as: China Youth Science and Technology Awards (2007), Youth Architect Awards of Architectural Society of China (2005), International Architects Salon Awards (2004), THE CHICAGO ATHENUM International Architecture Awards (2010/2009), Gold/Silver Prize of China National Outstanding Architecture Awards (2009/2010/2000), World Youth Chinese Architects Awards (2007), China Architecture Art Awards (2003), and Finalist of World Architecture Awards of UK (2002). He was also invited to take part in some exhibitions concerning architecture and art, such as "From Beijing to London: 16 Contemporary Chinese Architects", London (2012), "Verso Est – New Chinese Architectural Landscape 2011", Rome (2011), "Chinese Regional Architecture in a Post-Experimental Age", Karlsruhe/Prague (2010), "Heart-Made – The Cutting-Edge of Chinese Contemporary Architecture", Brussels (2009), 11th Venice Biennale of Architecture (2008), "Illusion Into Reality: Chinese Gardens for Living", Dresden (2008), "Get It Louder", Beijing (2007), "Happen", Left & Right Art Zone, Beijing (2007), 1st/2nd Shenzhen Biennale of Urbanism & Architecture (2005/2007), and "Status": Eight Young Chinese Architects, Beijing (2005).

Atelier Li Xinggang advocates the process of "Research + Design", and emphasizes the reaction of architecture to context, nature, site, material, user. They support the full-process of design concerning control of the building, interiors, and landscape. Thus they approach contemporary architectural works with cultural depth and aesthetic affection. The atelier also launches its studies on "Chinese Gardens – Settlements", "Reading Masters", and so on.

李兴钢工作室,2011,摄影:黄源 / Atelier Li Xinggang, 2011, Photograph: Huang Yuan